BEACH POLITICS

Beach Politics

Social, Racial, and Environmental Injustice on the Shoreline

Edited by

Setha Low

NEW YORK UNIVERSITY PRESS

New York, New York

NEW YORK UNIVERSITY PRESS
New York
www.nyupress.org

Library of Congress Cataloging-in-Publication Data
Names: Low, Setha M., editor.
Title: Beach politics : social, racial, and environmental injustice on the shoreline /
edited by Setha Low.
Description: New York : New York University Press, 2025. |
Includes bibliographical references and index.
Identifiers: LCCN 2024009082 (print) | LCCN 2024009083 (ebook) |
ISBN 9781479821945 (hardback) | ISBN 9781479821952 (paperback) |
ISBN 9781479821969 (ebook) | ISBN 9781479821983 (ebook other)
Subjects: LCSH: Beaches—Political aspects. | Beaches—Social aspects. |
Beaches—Environmental aspects. | Environmental justice. | Social justice.
Classification: LCC GB451.2 .B399 2025 (print) | LCC GB451.2 (ebook) | DDC
304.20914/6—dc23/eng/20240922
LC record available at https://lccn.loc.gov/2024009082
LC ebook record available at https://lccn.loc.gov/2024009083

This book is printed on acid-free paper, and its binding materials are chosen for strength and durability. We strive to use environmentally responsible suppliers and materials to the greatest extent possible in publishing our books.

Manufactured in the United States of America

10 9 8 7 6 5 4 3 2 1

Also available as an ebook

To my mother, Marilyn Rudley, who loved the beach,
and to my sister, Anna Harwin, for her memory
and kindness

CONTENTS

LIST OF FIGURES

PREFACE

SETHA LOW

This book began in 2003 when my mother sent a *Los Angeles Times* article about the restriction of public access to Broad Beach in Malibu "where non-resident sunbathers, picnickers and others are booted off the dry sand, which the community considers private property."[1] California's public access law guarantees beach access only seaward of the mean high tide line, the portion of the beach with damp sand. But many oceanfront communities—including Broad Beach—granted public easements to an additional twenty-five-foot strip of dry sand with access corridors. Nonetheless, Broad Beach is littered with no trespassing signs to keep the people away from the homes of celebrities such as Steven Spielberg, Danny DeVito, Goldie Hawn, and Dustin Hoffman. These wealthy homeowners assess themselves $3,000 to $5,000 per year for security guards to maintain their privacy, and two homeowners obstructed the legal public easement that ran between their houses.[2]

In August 2003, California coastal commissioner Sara Wan and a group of beach activists went to Broad Beach and refused to budge when confronted by a security guard on an ATV. The guard called the Sheriff's Department, and when five officers arrived, Wan explained that she was sitting on a public beach. She produced documents showing that despite the no trespassing signs, the homeowner had granted public access twenty-two years ago in exchange for the expansion of their home.[3] Environmental activist Jenny Price has since created an app, "Our Malibu Beaches," that identifies bogus signs that visitors can ignore.[4]

Ten years later, a dispute began over the use of fire rings that dot the Newport Beach shoreline in Orange County, Southern California. Fire rings (a regulation fire pit used in California) were built by beach goers for over sixty years, but beachfront residents started to complain about the smoke. The residents argued that open bonfires pollute the air and lead to many health problems, calling on the city manager to remove them. According to beach goers, concern over the health consequences of fire pits was just another excuse invented by homeowners to keep people off public beaches.[5] The *Orange County Weekly* editor was quoted as saying "It's the elite who live in these affluent communities. They don't want the hoi polloi to come down here."[6]

The South Coast Air Quality Management District that monitors air quality agreed with the residents and introduced a proposal to ban fire rings throughout Southern California. The Orange County Board of Supervisors voted to formally oppose the ban but supported the right of local municipalities, such as Newport Beach, to decide their own future. In the end, the California Coastal Commission was asked to remove fire rings in several well-to-do beach communities including Newport Beach, Balboa, and Corona del Mar.[7]

Meanwhile, on the eastern shore of Long Island, New York, where I spend summers, a private community—also known for its Hollywood entourage—drained East Hampton Town's Georgica Pond because their basements were flooded from heavy rains.[8] There is a long history of antipathy between the townspeople, who use the public pond to supplement their diet with fish, crabs, and clams and enjoy weekend sailing and canoeing, and the homeowners, who find the pond's high water table an annoyance. Normally a trench was dug from the pond to the ocean in the fall to improve the pond's water quality. But Georgica Associates, the community adjacent to the pond, wanted to reduce the water level to dry out their basements in the spring. The elected trustees in charge of natural resources denied their request, stating that the pond was for everyone's enjoyment. Nevertheless, a trench was cut that emptied the pond for the entire summer. The police gave up searching for the person who did the trenching, and no one has ever been charged.[9]

The most recent conflict on the eastern end of Long Island involves an injunction against pickup trucks driving on Napeague Beach, a popular strip of sand where professional and recreational fishermen have been catching striped bass, bluefish, and other species for generations. The residents of a condominium constructed along the Napeague dunes in 2015 imposed an injunction that challenges the right of fishermen to drive there, even though beach access was granted in 1801.[10] These struggles over beach access for fishermen and local recreation are more frequent with the increased value of ocean- and pond-front property in this region.

Beaches and ponds are significant public spaces for residents of Southern California and the East End of Long Island—as they are for those who live in many communities along the water. However, illiberal rules and regulations and coastal management manipulation augment illegal actions such as unauthorized no trespassing signs, fences, and security guards to enclose and restrict beaches so they can't be used or accessed. Strikingly, these exclusionary practices are bolstered by local and state governments who back homeowners rather than the general public to retain high property values that produce tax revenues and contributions to elected officials' political campaigns.

Even these dubious strategies, however, can be inadequate when dealing with large tracts of land surrounding a pond, lake, or beach. In such cases, limiting access through conservation easements or buying up private property encircling the beach or lake keeps the public out and contains its use. For example, large tracts of Colorado wilderness have been purchased by rich fly-fishing aficionados to retain the best rivers and streams for themselves.[11]

Building walls and fences with guard stations to barricade the beach is another means of excluding people. In Narragansett, Rhode Island, limited parking space and high parking fees did not reduce the crowds of nonresident beach visitors. So, the town added a chain-linked fence with gates where guards sell identification bracelets to allow entry. Charging $15 for parking and $12 per bracelet successfully deters nonresidents from using the beach.[12]

These physical impediments are bolstered by legal and financial institutions that place public land, such as a beach or lakeside, in the hands of a private corporation or a public-private agency. Legal schemes are harder to identify as they utilize normative governance procedures to ensure resident-only beach use. In Naples, Florida, residential communities with homeowners' associations (HOAs) annex beaches by building facilities and restaurants that can only be used by residents with passes and accessed by ferries.[13]

During this same twenty-year period, Dana Taplin, Suzanne Scheld, and I, along with graduate student members of the Public Space Research Group, completed rapid ethnographic assessments (REAPs) of parks, some with beaches. We investigated urban beaches such as Jacob Riis in Queens and Orchard Beach in the Bronx, New York City; suburban beaches such Jones Beach on Long Island, New York; and the lakeside beach at Lake Welch in Bear Mountain, New York. This research provided insights into conflicts about state and local governance, cultural preferences, and community context. Because these beaches are managed by the National Park Service (Jacob Riis), New York City (Orchard Beach), and New York State (Jones Beach and Lake Welch), we learned about the value of beaches as public places where people feel they belong and are respected—regardless of immigration status, cultural identity, race, ethnicity, language, age, or gender.

Fieldwork at contested beach sites in Southern California; Long Island, New York; Rhode Island; and Florida, on the other hand, uncovered numerous forms of exclusionary and racist tactics employed to restrict their use. These findings suggested a much broader range of unjust practices could be found in the United States, and site visits to beaches in Costa Rica, Mexico, Slovenia, Croatia, and South Africa, suggested they could be found globally. Thus, while this book began with anger and activism against illegal and illiberal strategies of beach exclusion in the United States, it evolved into an exploration of how beaches are fraught with contested claims and sociopolitical inequality worldwide.

NOTES

1 Kenneth R. Weiss, "A Malibu Civics Lesson: Beach is Open," *Los Angeles Times*, August 25, 2003, www.latimes.com.
2 M. Navarro, "In Malibu, the Water is Fine (So Don't Come In)," *New York Times*, June 5, 2005, ProQuest.
3 Weiss, "Malibu Civics Lesson."
4 Dougal Henken, "America Has a Private-Beach Problem," *The Atlantic*, September 12, 2023, www.theatlantic.com.
5 Noel King, "Beach Fire Pits Spark Battle Between Private Interests and Public Land," *NPR, Marketplace*, April 8, 2013, www.marketplace.org.
6 King, "Beach Fire Pits Spark Battle."
7 Jill Cowan, "Orange County Supervisors to Weigh in on Beach Fire Ring Ban," *Los Angeles Times*, April 22, 2013, www.latimes.com.
8 Ken Moran, "Who Opened Georgica Pond?," *New York Post*, July 9, 2003. https://nypost.com.

9   Michael Shnayerson, "Another Hamptons Whodunit: Who Drained East Hampton's Georgia Pond?," *Vanity Fair*, March 11, 2015, www.vanityfair.com.

10 Christopher Walsh, "Truck Beach Trial Ends with Swords Drawn," *East Hampton Star*, June 16, 2016, 1; Christopher Walsh, "Court Says No Driving on Napeague Beach," *East Hampton Star*, June 16, 2021, 1; Christopher Walsh, "Judge Orders 6K Beach-Driving Permits Revoked," *East Hampton Star,* July 7, 2022, 1.

11 Ben Ryder Howe, "Unspoiled Waterways, Unavailable to Some," *New York Times*, September 4, 2022, 6.

12 Setha Low, fieldwork notes from Narragansett, Rhode Island, 2023.

13 Setha Low, fieldwork notes from Pelican Bay, Naples, Florida, 2021.

Introduction

SETHA LOW

Visitors to Jones Beach, a majestic state park on Long Island, New York, talk about how important it is that the beach is public because there is always a place for them. They reflect on how being together on the sand or boardwalk and in the water creates a sense of solidarity, cohesion, and belonging.[1] That is what a beach can offer if it is public.

The intention of this book is to provide evidence of what happens when the beach is not public and launch an inquiry into why the public is losing access to and control of public beaches and their valuable common resources. The contributors were asked to draw from their research experience to produce case studies that uncover and illuminate the political goals, social relationships, and economic intentions whenever beaches are lost, redeveloped, or restricted. The explicit goal is to generate a body of empirical evidence on the politics of these disputes and encounters.

Beaches—from seashores to lakesides and riverfronts—are a unique form of public space valued for their environmental, social, cultural, aesthetic, spiritual, and affective resources and meanings. They are places where water and land come together creating shorelines of opportunities for working, playing, relaxing, strolling, socializing, picnicking, and living—in scenic seaside homes that range from self-built shelters to multimillion-dollar mansions. Beaches are so precious they become contested terrains where people fight to control their assets and capitalize on their economic potential rather than retain them for public use. Ecologically fragile because of constant change due to tides, storms, and ocean warming adds a geophysical dimension to the already complex interactions of the beach's multilayered dynamics. Coastal deterioration and intensifying storms, especially hurricanes and flooding, appear in these chapters, but geomorphology is secondary to an emphasis on politicization, racialization, and commodification. Nevertheless, the study of beach politics provides clues to the critical need for climate change adaptation.

In this volume, we approach the beach as not just "a depiction of a sandy and tropical strip of nature next to blue waves," but a social construct and category of social experience, a place of everyday life and difference, a contested commons, and a collective imaginary.[2] It provides a staging ground and temporal-spatial edge that illuminates community solidarity and protest in the face of property loss, racial erasure, class conflict, and spatial governmentality. Materially, the beach is an ever-shifting geomorphological formation, an unsettled borderland between land and water, and for many, a beloved strip of sand. As an ethnographic and historical place, it becomes a landscape and moral terrain on which struggles for control, ownership, and rights are enacted daily—a "contact zone" where race, class, gender, age, ability, and identity encounters are on display. Community battles against developers and the often-ambivalent role of governments with different jurisdictional responsibilities become visible through the mediation and negotiation of beach access and use rights. Hidden stories, forgotten memories, illegal actions, and exclusionary legislation transform a "strip of nature" into a palimpsest of greed, racism, ecological disregard, and socioeconomic discrimination.

Personally, as I recount in the preface, I am concerned with the illegal and illiberal ways that beaches are restricted in the places I have lived. Theoretically, the book is guided by previous publications on the politics of public space and the production of security spaces that reveal the entanglement of histories, infrastructures, socio-spatial relations, and affects that shape urban places.[3] For instance, multiple mechanisms of securitization—spatial governance through enclosure and surveillance; political governance through laws, regulations, and policing; economic and financial control through security capitalism; and affective atmospheres of fear and anxiety—are necessary to produce "securityscapes."[4] Beaches are similarly embedded in a tight knot of spatial governmentality (enclosure), political regulation (legislation, access rules, and monitoring), economic control (property ownership and land development), and affective attachments (memories, social relations, and everyday activities). Beaches are also governed by environmental and ecological processes and a distinctive moral order based on dress and demeanor.

What do confrontations over beach use—the various restrictions, justifications, and access conflicts—tell us about contemporary politics? They remind us that coastlines—especially their most coveted parts, sandy beaches—are a reflection, and refraction, of contemporary political and economic struggles.[5] In the most general terms, this book focuses on these struggles especially as they are grounded in a particular time and space—that is, beach "politics."

Politics, at a conceptual level, are what anthropologists Nancy Postero and Eli Elinoff call the "practices of world-making that proceed through the formulation of constellations of critique, disagreement, difference, and conflict" and are a part of a struggle to live "otherwise."[6] In their theoretical analysis, everyday politics are embedded in the phenomenology of place-making and embodied space and encourage a lived-experience lens be applied to those who are seeking to remake their worlds.[7] Some contributors employ an embodied and experiential framing through vignettes and interviews with people who talk about why they are involved in fighting for (or against) a beach intervention. These renderings offer a glimpse of the world-making and collective imagining that is part of ongoing beach politics.

French philosopher Jacques Rancière, in contrast, theorizes politics as an emergent process that occurs as actors reconstitute the political community by asserting their claims as legitimate political beings.[8] He argues that "politics exists when the natural order of domination is interrupted" and notes that this occurs especially when the argument is about the distribution of common lots or communal shares such as the use of a beach.[9] "Politics" according to Rancière is a rupture in the normal distribution of those who exercise power and those who are subject to it.[10] If the principal function of politics is to disclose the world and its subjects as made manifest through dissent, then beach conflicts are indeed "politics" in that they reveal social inequalities at play.[11] Both of these theoretical framings are useful for thinking about beaches as the basis of political transformation, confrontation, and democratic change.

One other theoretical approach to politics is American pragmatist John Dewey's concern with the dynamics of publicization and his admonition that publicity (making something public) requires an inquiry into an unsettled or disturbed situation.[12] Sociologist Cédric Terzi and environmental psychologist Stéphane Tonnelat take Dewey's ideas and apply them to the study of public space, arguing that for a space to become truly public it must be "troubled" to draw attention as a problem to be solved. A problem must be seen to be addressed, or in their terms, "publicized," and it is through this publicization that a politics of public space emerges.[13] This framing seems particularly useful to the beach conflicts and disputes presented. Although not explicitly referred to in any of the studies, the unsettled shores of beaches offer the visibility and troubles suggested by this theoretical framing.

Over the past fifty years, the privatization and development of public beaches has accelerated with the help of local elites, governmental interventions, and international corporations. Despite this, beaches have not been a focus of study from a public space perspective. What we do know is that when disputes arise, the powers that be often side with private interests and wealthy community members.

As a result, only small portions of the shoreline are available to the public regardless of their legal designation.[14] In most of the United States, for example, the Public Trust Doctrine, a property law from Roman times that became English and colonial US common law, ensures that all tidelands and lands under navigable waters are owned by the state for the benefit of all citizens with certain rights of usage. Private owners are also under an obligation to protect the public interest and not interfere with the public's rights.[15] But every state has its own history of defining what constitutes the "tidelands," resulting in a patchwork of regulations for what "holding in trust for the enjoyment of public rights" means. The National Open Beaches Act of 1969 was intended to outlaw "any obstruction, barrier, or restraint of any nature which interferes with the free and unrestricted right of the public . . . to enter, leave, cross or use as a common the public beaches." But it never passed in US Congress due to arguments that it would lead to overuse—the "tragedy of the commons" argument of the ecologist Garrett Hardin.[16] Hardin argued that if a valuable resource, like a beach, was left for people to use freely it would be overused and its value destroyed.[17]

The Public Trust Doctrine—and its evolution and defense—is only one facet of beach dispossession and exclusion in the United States. To understand the significance of local conflicts requires a broader investigation into the circumstances in which use rights and ownership are contested. In the United States, racial and class dynamics come into play, as it is often white, wealthy property owners who disregard the law and restrict access to waterfront property regardless of public legal protections.

Similar kinds of duplicity and corruption that disregard national and state laws guaranteeing public access exist globally albeit in other forms and guises. Neoliberal strategies that commodify beach property are employed by private companies and the state in Europe, the Middle East, and Latin America. To outline the contours of beach politics more broadly, then, the chapters document a variety of national and state contexts, from Brazil and Argentina in Latin America; Puerto Rico in the Caribbean; South Africa in Africa; Greece, Austria, France, Ireland, and Germany in Europe; Lebanon in the Middle East; and Australia. Within the United States, case studies from California, Connecticut, and New York are included with an emphasis on New York City and Long Island cases.

The contributors to this book, known for their work on public space and social justice, showcase how public beaches are inextricably bound with spatial segregation and enclosure of the commons. As this cross-cultural comparison makes clear, even when shorelines and coastal areas are legally accessible, there remain ways in which people are excluded, dislocated, or dispossessed. Whether vendors in beach towns in Argentina; Black residents in Salvador, Brazil; beach goers in Sydney, Australia; or lake swimmers in the mountains of Northern Austria, there are multiple and widespread barriers to accessibility of the shoreline for all.

Scholars in the humanities, social sciences, and design and planning professions come together in this volume to produce significant resonances and creative synergies that enhance the breadth of the project. Social scientists offer a range of methodologies including long-term ethnographic fieldwork in Brazil, Argentina, and Puerto Rico; auto-ethnography and activist ethnography in New York City; rapid ethnographic assessment procedures (REAP) in Fire Island, Long Island; and multisited ethnography in Mexico and California. Historians utilize archival materials, legislative records, oral histories, expert interviews, and media sources to reconstruct beach use in Connecticut and Dublin. Forgotten photographs and museum exhibitions visually recreate a lakeshore in Albuquerque, New Mexico. Socio-spatial planning analyses in Beirut and mixed methods in Australia, Greece, and Austria document land use and institutional changes. The studies reveal interconnections in their findings and interpretations despite differing epistemological assumptions. The range of disciplines and methods adds to the book's usefulness and methodological richness.

* * *

This book examines beach politics at two levels: (1) specific beach conflicts occurring at distinct times and places, and (2) more general explorations of the underlying social structures and political institutions. It queries the processes and practices involved in the public's loss of the beach—especially for marginalized peoples. What are the circumstances that allow the enclosure of the beach and restriction of access to what was once a commons? Where and in what situations do these exclusionary strategies occur? How is the denial of public rights justified and rationalized—morally and legally? Contributors address these questions by uncovering contested claims and inequalities that exist in each cultural setting. By examining challenges to existing power structures, they reveal solidarities that resist, repair, and restructure injustice.[18]

The chapters are organized in four parts that correspond to the most frequent strategies and rationales used to deter public beach use in the case studies: (1) governing the beach, (2) shoring up the coastline, (3) racializing the beach, and (4) developing the beach. Each part places a selection of cross-cultural case studies in conversation with one another within the four overarching domains. While these domains overlap and are not mutually exclusive—indeed, many of them are referenced in the same chapter—they provide a useful framework to organize the dialogue and critique.

Part I highlights how policies, laws, regulations, and formal and informal governance systems restrict beach rights often resulting in protest and activism. Part II demonstrates how hardening or changing the water's edge limits public beach and water access leading to various forms of resistance. Part III focuses on visual representations, discourses, imaginaries, and policing of racialized bodies to maintain the beach as a white public space. Part IV characterizes how tourist development restructures who can use the beach, and how commodification takes beaches into the city center where they take on new meanings.

Part I. Governing the Beach: Policies and Protest

Beaches that were initially considered a commons that everyone could use, or an ignored backwater where poor people lived, changed dramatically in the United States beginning in the 1840s and 1850s, rapidly becoming vacation spots and valuable property throughout the twentieth century. With this increase in economic value, public beach lands were parceled and sold by developers especially along the shoreline of the northeastern United States. In the southern United States, many beach communities were made up of African Americans and Indigenous peoples who moved to what were vacant and marginal coastlands to farm, collect, or fish and build modest homesteads. Their lands were taken—often illegally—with little to no compensation to be used by the state or sold for private development.[19] A case in point is Bruce's Beach, owned by an African American couple who built a beach resort in Manhattan Beach, California. Their property was seized by eminent domain in 1924, and finally returned to the Bruce family ninety-five years later.[20] The extensive privatization and enclosure of beach land was facilitated by inventive legal instruments and collective ownership regimes that circumvented the rights and public claims laid out in the Public Trust Doctrine.

In European, Middle Eastern, and Latin American countries, where democratic governments ensure the public's right of beach access, other governance strategies such as master plans and zoning controls, development projects on beaches as amenities for high-end housing, and publicly owned management agencies empowered to sell shoreline property limit public use. In some cases, informal governance negotiations based on visibility, moral values, and social sanctions achieve economic and political goals. In the following cases, attempts at governing the beach sparked demonstrations, unexpected solidarities, and ongoing protests to stop the interventions—sometimes succeeding.

The historian Andrew Kahrl opens with a discussion of the US state institutions and legal mechanisms that enabled white homeowners along the Connecticut shoreline to privatize the beachfront.[21] His history of beach associations in the northeastern United States illuminates how homeowners' demands for exclusivity evolved from deed restrictions against selling to Jewish or Black people and from zoning requirements that concealed their underlying political and racist motives.[22] Beach association charters contained prohibitions against "nuisances," including the presence of nonmembers on land held in common, and the beach. Restricting who could buy property expanded to who could access, swim, walk, or relax there. Residents-only beaches were bolstered by racial discrimination in housing sales and by real estate agents, who refused to show homes to Black buyers. Ultimately the Connecticut Supreme Court ruled that towns could not bar access, but other modes of social ordering continue to control public beach use today.[23]

Contemporary disputes over governance also highlight the beach as a politicized terrain where constituencies with diverse interests come together. Anthropologist Matilde Córdoba Azcárate suggests identifying the cultural, scientific, and Indigenous repertoires used to justify rights to access, belonging, mobility, or environmental protection to show how disparate communities can become politically aligned. She focuses on protests generated by a 2018 attempt to fence part of the railroad tracks that give residents illegal access to the beach in Del Mar, Southern California. The transit district was under an obligation to protect the public from high-speed trains that run along the shore, but wealthy residents, environmental conservationists, social justice activists, as well as the California Coastal Commission, and other nonprofit and volunteer associations joined together to contest this infrastructural change.

The seafront spaces in Beirut, Lebanon, are experiencing pressures from tourism and gentrification but also from increasing pollution and loss of heritage in the face of ongoing civil war, governmental collapse, protest, and disorder. Nadine Khayat and Clare Rishbeth, a landscape planner and a landscape architect, respectively, address the increasing fragmentation of access to the beach, one of the few places that older residents and Palestinian and Syrian refugees traditionally use for leisure time and recreation. Based on the voices of users from diverse backgrounds and a range of stakeholder insights, they characterize the governance mechanisms used to gentrify adjoining neighborhoods and beachfront and at the same time abandon the remaining public beach that provoked local demonstrations and protests.

Anthropologist Mariano Perelman examines informal forms of governance on the coast of Argentina by interviewing vendors who strive to self-govern and escape "mafia" control and by following a conflict over two female beach goers' right to be topless. He employs these divergent ethnographic cases to argue that a particular kind of spatial-temporal configuration emerges in coastal towns during the summer season. This "beach time" is characterized by changes in the moral order and politically amplified by the visibility of bodies—of workers in the first case and topless sunbathers in the second. While the vendors informally manage their self-presentation through spatial restrictions and negotiations to make a living, the topless sunbathing conflict escalated through media attention and became a national protest for gender entitlement and political equity.

Urban studies scholar Sabine Knierbein and architect and urban planner Charis Christodoulou draw on two cases to take a more theoretical approach. Privatization of public beaches and lakeshores is an issue in the Mediterranean (Greece) and in Central Europe (Austria) even though the right to public waters is part of European Union (EU) policy. They view this transformation from a post-political perspective based on the increasingly difficult public access and wider depoliticization of basic rights to enjoy nature. In these EU states, economic development is enabled through EU and national policies and through depoliticized governance arrangements to develop and exploit the beach. The Greek and Austrian cases illustrate how common imaginaries have been appropriated by capitalist interests, facilitated through a post-political condition undermining basic public access rights to enjoy the beach as a public good. This is a defining characteristic of the post-political beach.

Part II. Shoring up the Coastline: Protection and Resistance

With global warming and climate change, there are devastating storms and the ever-present danger of rising water and loss of land. Especially on beaches made by dredging and shoring up the shoreline with sand, the impeding threat of their disappearing completely over the next few decades is conceivable. Risk assessments and cost analyses of the vulnerability of beaches and housing located adjacent to inundation areas are increasing at such a rate that many homeowners and communities can no longer obtain flood or storm insurance. Protecting the shoreline and its adjacent parks, houses, beaches, and other amenities from water damage and flooding is often the rationale for governmental and private decisions about hardening or building bulwarks based on the economic value and private property rights to be defended. Even while one might agree with the good intentions of protecting a valuable beach front, restoring a boardwalk, or building a jetty or wall to protect and retain the sand, the impact of saving a residence or protecting the shoreline often results in privileging those who own the land and not those who have the right to use the beach. The public loses their rights of access and ability to determine future use because "experts," often backed by private funding, determine how the environment should be protected.

Geographers Kurt Iveson and Ana Vila-Concejo examine the impact of the 2016 storm that eroded Collaroy-Narrabeen, the most capitalized beachfront in Sydney, Australia, damaging ten homes and collapsing a swimming pool. They address whether the protection of the public beach or of beachfront property should be the goal of public policy as such storms become more frequent in a changing climate. What is unusual about the Sydney case is that historically the local government had gradually begun to buy back at-risk private properties so that they could be removed and sand dunes restored. But Collaroy-Narrabeen property owners successfully pushed for a different resolution, one that involved the construction of a privately funded, state-approved seven-meter-tall seawall. The publicness of the beach was compromised by this deferential treatment of private property in the city's environmental adaptation strategies.

On Long Island, New York, tensions over how to protect a fragile ecology and the eighteen seasonal beach communities at Fire Island National Seashore plague the National Park Service (NPS) administration and local rangers. Protecting the shoreline with its sand migration, dune erosion, and breaches due to storms in a manner that serves summer residents' requirements of stable ground and property lines is antithetical to the NPS mission. The National Seashore's mandate to maintain and interpret the natural cycles of the barrier island system comes into direct conflict with residents and property owners who want to stabilize these forces. Environmental psychologist Dana Taplin provides a unique glimpse into the backstory of this contested space through ethnographic interviews with NPS employees who try to fulfill their institutional goals, even taking on their own environmental campaigns, and with residents who tell stories of the old days and believe they know what is best for the island.

After Superstorm Sandy, New York City opened a competition to redesign East River Park to withstand rising sea levels due to climate change. Lower East Side community gardeners, public space supporters, artists, designers, academics, environmental scientists, and some twenty-six community groups came together to initiate a dialogue. The result was the creation of a "consensus plan" that would preserve the park and prepare for flooding. Late in the process, the city rejected the consensus plan and favored a 2.4-mile seawall that would raise the park and pave it with artificial turf for ballfields, removing the trees and community space. Social worker Benjamin Heim Shepard's recounting of the grassroots fight to retain the consensus plan illustrates how the imposition of environmental engineering strategies to protect the Lower East Side excluded current users and local community participation and concerns.

Part III. Racializing the Beach: Inequality and Erasure

Racialization of the beach occurs through forced removal, discursive strategies, collective property regimes, and representational erasure. Unlike environmental or economic justifications, defending whiteness is only acceptable when cloaked in layers of obfuscation, disbelief, and misdirection. It is hard to characterize the beach politics of the United States, South Africa, and Brazil without discussing the racism that underlies whatever strategy is used to restrict entry, ownership, or use. The privileging of whiteness and the erasure of Black bodies is particularly salient for understanding how racialization is produced but also struggled against by grassroots activists and home communities.

The impact of Hurricane Sandy in 2012 generated a cascade of procedures to protect local beaches, but these strategies of storm mitigation and beach redevelopment also reinforced a "racial coastal formation." Drawing upon long-term research on Rockaway, Queens, environmental psychologist Bryce DuBois and urban planner Lee Graham reveal how the post-Sandy Special Initiative for Rebuilding and Resiliency (SIRR) continued the displacement of predominantly Black residents that began in the 1950s–1960s due to urban renewal. The SIRR emphasized getting New Yorkers back to the beach as a symbol of the city bouncing back and focused on rebuilding the boardwalk. These benefits did not extend eastward, where Black residents suffered from the inundation from Jamaica Bay that filled basements with water, damaged housing, and produced disease-causing mold and contamination. The emphasis on the boardwalk ignored the damaged living spaces of Black residents in the adjacent low-income and public housing.

Geographer Natasha Howard explores another avenue for understanding how recreational space was racialized and white privilege preserved in Albuquerque, New Mexico. By analyzing photographs of Tingley Beach, a human-made lake popular from the 1930s through the 1950s, she uncovers the formation of a white spatial imagery present in a contemporary museum exhibit, the city's website, and the Albuquerque Museum photo archive. She argues that the photographs are a socially constructed history that memorializes a singularly white representation of the beach. The rich archive of photos, film, and other ephemera offer no documentation of Black visitors, even though there was a Black community not far from the beach. Howard theorizes that claims to representation are spatial claims to belonging. Not acknowledging the ways whiteness shaped the spatial history of the beach is the work of racial erasure, another kind of dispossession based on representational amnesia and collective memory distortions.

Social psychologist Kevin Durrheim traces the history of white South African beach goers through four stages of defending settler whiteness in Durban, South Africa. He argues that whiteness is strongly tied to privilege that bestows advantages on white-classified people. As beaches became spaces of vacationing and leisure, they also became an important symbolic anchor for white entitlement. With the end of apartheid, the racial hegemony and segregation of the beach crumbled, challenging white people's sense of worthiness and eliciting rage and shame. Durrheim studies this historical transition through the discursive strategies employed to express and defend white superiority and political subjectivity.

Anthropologist Keisha-Khan Perry's ethnography of the dispossession of Black coastal communities in Salvador, Brazil, pivots between the terror of forcible removal of beachside communities to make way for luxury apartments and restaurants and the solidarity of Gamboa activists fighting to retain their homes. She uncovers discursive strategies in this struggle, observing that initially, Gamboa de Baixo families did not know that they lived in "a coveted paradise" on the coastlands of the Bay of All Saints. Once the beach lands became valuable, residents began to take their "paradise" more seriously, battling developers and police to resist displacement. A similar "right to paradise" struggle is underway along the Garifuna coast in Honduras, where claims of Indigenous Blackness are the basis of defending land and autonomy.[24]

Part IV. Developing the Beach: Tourism, Activism, and Regulation

Promoting urban development and tourism through the commodification of beach land to increase its economic value for national and global projects is a central theme of the beach literature. Matilde Córdoba Azcárate's ethnography, *Stuck with Tourism*, Waleed Hazbun's account, *Beaches, Ruins, and Resorts*, as well as Sarah Stodola's personal chronicle, *The Last Resort*, document the social and ecological devastation of beach development on local cultures and peoples.[25] Contemporary studies of the private takeover of local beaches for tourism in Puerto Rico; sporting activities such as golf courses, rifle ranges, horse races, and hunting in Dublin, Ireland; and the production of artificial beaches across Europe emphasize the solidarities and growing political awareness occasioned by these development activities with differing outcomes.

Anthropologist Katherine McCaffrey focuses on sovereignty and opposition to unchecked development in her study of beach conflicts and protests in Puerto Rico. She draws on her previous research on Vieques Island's gentrification and military displacement to focus on the relentless transformation of the beachfront into exclusive hotels and tourist destinations.[26] The dream of living along the shore, collecting seashells, and fishing, is replaced with a multipronged fight for self-governance and a sustainable future. McCaffrey ends her story by highlighting twenty-five years of confrontations between environmental activists, politicians, and various hospitality companies including Marriott, the Normandie Hotel, and the Paseo Caribe over the development rights to coastal beach lands.

Historian Paul Rouse writes that Dollymount Strand and Bull Island on Dublin Bay historically offered urban dwellers a place of leisure and recreation because of their proximity to the city center. The unique location and habitat were initially used for sea swimming and as a seaside resort but later became a place for organized sports. With the establishment of golf as a sporting world, wealthy landowners bought existing royal land to construct two golf courses in the dunes, altering the physical landscape and installing fencing that marked it as private property. With Bull Island's designation as a UNESCO biosphere reserve in 1981, the tension between its ecological and leisure uses generated enough public protest that in 1994 the Dublin City Council declared that no development of any significance could be undertaken without permission except for the two privately deeded golf courses. But even with this protection, there remains tension between the beach publics—nature lovers, sports enthusiasts, and pleasure seekers—who continue to have conflicts about appropriate beach use today.

The value of beaches has not gone unnoticed by governments and private entrepreneurs who see the benefit of having beaches available in cities as an alternative form of recreational and consumption space. Urban designer Quentin Stevens describes an array of artificial beaches in Germany and France and examines how they are developed, their problems and politics, and whether they offer the various kinds of rights that define spatial control. Stevens points out that the flexibility, limited time frame, and ease of mobility of artificial beaches means they can occupy urban edges and leftover or deteriorating spaces that otherwise would be overlooked, adding value where there was none. But at the same time, artificial beaches can be used to loosen planning controls or become part of a waterfront redevelopment strategy for upmarket housing to spur further gentrification. Waterfront parks and beaches in the United States such as at Hudson River Park in Manhattan, Brooklyn Bridge Park in Brooklyn, and the just-opened riverfront park in Memphis, Tennessee, are examples of this shoreline development strategy.[27]

Conclusion

Beaches and other waterfronts are beloved by people for their recreational, environmental, and work opportunities. Indeed, hanging out, fishing, and collecting along the shore of a body of water are some of the most ubiquitous human activities. Yet, as this volume documents, in communities across the United States and around the world, there is an alarming trend of restricting access to public sections of the beach to ensure that waterfront property owners or tourists are the only ones able to access the shoreline.

Several cross-cutting themes emerge from the collective research findings.[28]

1. Community conflicts and interventions are often the result of coastal engineering and urban development that reshape the social relationships of key actors and the socio-racialized space of beaches.
2. Emergent moral economies and moral orders are created through beach conflicts and disputes co-opting the existing system and reframing it in such a way that beach exclusivity and exclusion are deemed acceptable and socially just.
3. Collective memories are represented and retained differentially depending on discursive, visual, and symbolic practices. When collective memories are repressed or erased, it reduces a sense of place attachment or belonging, enabling beach loss—both materially and metaphorically.
4. Law and policy play a critical role in the structuring of what is perceived as crime and the process of criminalization. The illegal activities of those in power are hidden or abetted by the legal and regulation system while protest activities and public use are criminalized.
5. Political action and discourse function in relation to existing laws and policies. Individuals, companies, and governments tactically implement, fail to implement, or break existing laws and policies by exploiting jurisdictional ambiguities.
6. The geophysical and material environment is a sociopolitical stage for the enactment of beach struggles but is also an actor to the degree to which the beach becomes valuable as an economic resource.
7. The transition of the beach from use value to exchange value within capitalist economies is a major factor in the transformation of the beach from a public space to a commodity and coveted property. In this transition, the beach is reconstituted from its origins as a commons with public rights to a regulated territory (property) with limited rights of use and access.

These themes suggest new research directions, theoretical perspectives, and focused publications. The hope is they provoke lively discussion and political action to save our beaches—and understand why they must be saved—for the future.

NOTES

1 Setha Low, *Why Public Space Matters* (New York: Oxford University Press, 2023).

2 Sabine Kneirbein and Charis Christodoulou, "The Post-Political Beach: Conceptual and Empirical Exploration in Greece and Austria," this volume.

3 Setha Low and Neil Smith, *The Politics of Public Space* (New York: Routledge, 2005); Setha Low, *Behind the Gates* (New York: Routledge, 2003); Setha Low and Mark Maguire, *Spaces of Security* (New York: New York University Press, 2019).

4 Mark Maguire and Setha Low, *Trapped* (Palo Alto: Stanford University Press, 2024); Low and Maguire, *Spaces of Security*.

5 Jennifer Bidet and Elsa Devienne, "Beaches of Contention," *Actes de la Recherche en Sciences Sociales* 218, no. 3 (2017): 4–9.

6 Nancy Postero and Eli Elinoff, "Introduction: A Return to Politics," *Anthropological Theory* 19, no. 1 (2019): 3–28.

7 For other embodied and queer understandings of embodied beach place making, see Jah Elyse Sayers, "Black Queer Times at Riis: Making Place in a Queer Afrofuturist Tense," *Wagadu: A Journal of Tansnational Women's and Gender Studies* 22 (2021): 57–104; Vanessa Agard-Jones, "What the Sand Remembers," *GLQ* 18 (2012): 2–3; Laura Junka, "Camping in the Third Space: Agency, Representation, and the Politics of Gaza Beach," *Public Culture* 18, no. 2 (2006): 349–59.

8 Postero and Elinoff, "Introduction," 9.

9 Jacques Rancière, *Disagreement: Politics and Philosophy* (Minneapolis: University of Minnesota Press, 1999), 11; Mark Davidson and Kurt Iveson, "Occupations, Mediations, Subjectificaitons: Fabricating Politics," *Space and Polity* 18, no. 2 (2015): 137–52.

10 Jacques Rancière, Davide Panagia, and Rachel Bowlby, "Ten Theses on Politics," *Theory & Event* 5, no. 3 (2001).

11 Rancière, Panagia, and Bowlby, "Ten Theses on Politics."

12 John Dewey, *The Public and Its Problems* (Athens, OH: Swallow Press, 1927); John Dewey, *Logic: The Theory of Inquiry* (New York: Holt, Rinehart, and Winston, 1938).

13 Cedric Terzi and Stephane Tonnelat, "The Publicization of Public Space," *Environment and Planning A* 49, no. 3 (2017): 519–36.

14 Michael Waters, "America Has a Private-Beach Problem," *The Atlantic*, September 12, 2023, 9.

15 Jack H. Archer, Donald L. Connors, Kenneth Laurence, Sarah Chapin Columbia, and Robert Bowen. *The Public Trust Doctrine and the Management of America's Coasts*. Amherst: University of Massachusetts Press. 1994.

16 Andrew Kahrl, *Free the Beaches: The Story of Ned Coll and the Battle for America's Most Exclusive Shoreline* (New Haven: Yale University Press, 2018).

17 The "overuse of the commons" argument is less frequently employed today, and private property claims have become more salient and important.

18 Twenty years ago, I found the rationalizations of why people live in gated communities helpful in uncovering the fear, racism, financial insecurity, and economic incentives behind their decisions.

19 Kahrl, *Free the Beaches*.

20 Clyde McGrady, "Sale of Beachfront Land Spurs Debated on Reparations Goals," *New York Times*, February 19, 2023, 1.

21 Andrew W. Kahrl, "Fear of an Open Beach: Public Rights and Private Interests in 1970s Coastal Connecticut," *Journal of American History* 102, no. 4 (September 2015): 433–62. This chapter draws on this article and Kahrl, *Free the Beaches*, for research conclusions and historical materials.

22 Andrew W. Kahrl, *The Land Was Ours: How Black Beaches Became White Wealth in the Coastal South* (Chapel Hill: University of North Carolina Press, 2012); Andrew W. Kahrl, "The Sunbelt's Sandy Foundation: Coastal Development and the Making of the Modern South," *Southern Cultures* 20, no. 3 (2014): 24–42.

23 Adam Keul, "The Fantasy of Access: Neoliberal Ordering of a Public Beach," *Political Geography* 48 (2015): 49–59.

24 Christopher A. Loperena, *The Ends of Paradise: Race, Extraction, and the Struggle for Black Life in Honduras* (Stanford, CA: Stanford University Press, 2023). See also Ulrich Oslender, *The Geographies of Social Movements: Afro-Colombian Mobilization and the Aquatic Space* (Durham, NC: Duke University Press, 2016).

25 Matilde Córdoba Azcárate, *Stuck with Tourism: Space, Power, and Labor in Contemporary Yucatan* (Berkeley: University of California Press, 2020); Waleed Hazbun, *Beaches, Ruins, Resorts: The Politics of Tourism in the Arab World* (Minneapolis: University of Minnesota Press, 2008); Sarah Stodola, *The Last Resort: A Chronicle of Paradise, Profit, and Peril at the Beach* (New York: Ecco, 2022).

26 Katherine McCaffrey, *Military Power and Popular Protest: the U.S. Navy in Vieques, Puerto Rico* (New Brunswick: Rutgers University Press, 2002).

27 Jane Margolies, "Shiny Towers, and Islands and a Beach, Too," *New York Times*, September 19, 2023, 5.

28 I would like to thank the manuscript reviewer for the insights discussed in the conclusion.

PART I

Governing the Beach

Policies and Protests

Mile Hill Beach Edge (Setha Low, 2023)

1

Enclosing the Coastal Commons

Beach Privatization and Social Inequality in the
Twentieth-Century United States

ANDREW W. KAHRL

"In recent years fences and barricades have blocked the public right to have access to our seas. We are becoming a landlocked people, fenced away from our own beautiful shores, unable to exercise the ancient right to enjoy our precious beaches." This was how Texas Senator Ralph Yarborough characterized the state of America's coastlines in 1969. Over the previous decades, residential and commercial development spread across vast stretches of the coastal United States. As it did, public access to beaches that, legally and traditionally, had been treated as a commons, open to all and owned by no one, dwindled as beachfront homeowners, industries, and commercial enterprises sought to claim the beach—and the enhanced values and capital-generating capacities it provided—for themselves.[1]

Over the course of the twentieth century, a massive land grab took place along America's coastlines, one that resulted in the wholesale expropriation of public lands and resources by private entities and interests. By some estimates, upward of 95 percent of the seashores in America that were suitable for public use had, by the late 1960s, been effectively closed off to the general public. This enclosure of America's coastal commons unfolded in piecemeal fashion, one beachfront lot, one local ordinance, and one newly emerging coastal real estate market at a time. But in sum, it constituted one of the most socially and environmentally consequential commodifications of a common resource and one of the most distinctive forms of accumulation by dispossession witnessed in modern US history.

This chapter tells this history and unpacks its broader implications for the study of social inequality and its environmental dimensions in the

twentieth-century United States. It does so through tracing the enclosure and privatization of the Connecticut coast. This 253-mile stretch of
shore underwent one of the most thorough and contested enclosures of
any in the United States. By the 1970s, save for a few scattered, and often
overcrowded, state beaches, the entirety of the state's coastline had either been appropriated by heavy industries or claimed as a private playground for wealthy beachfront homeowners and residents of exclusive
shoreline towns. The legal and policy instruments used to enclose and
privatize Connecticut's shoreline, the interests and motives behind this
enclosure, and the social and environmental repercussions that resulted
offer valuable insights and context for understanding beach politics in
the twentieth-century United States more generally and for situating histories of beaches and coastal environments within the larger currents of
modern US social and political history.

* * *

Legally, beaches are the people's property. The Public Trust Doctrine
defines the foreshore as public land, a legal principle that dates back to
the Roman era and was incorporated into American jurisprudence from
English Common Law at the founding of the republic.[2] Each state, however, marked the line separating public land from private property along
the shore at a different spot—some drew the line at high tide, others at
low tide, still others at the vegetation line—and devised different defini-
tions of what constituted legitimate use of the public's shore.

Throughout the nineteenth century, beaches' legal status as public land was reflected in social patterns of use. With the exception of
heavily developed urban waterfronts, the lands that hugged the nation's
shorelines were sparsely developed and treated as a commons. Building
houses by the sea was widely—and wisely—understood to be a risky
proposition. Permanent structures were vulnerable to the ravages of
storms and recurrent changes in the shape of the shoreline itself, which
exist in a state of dynamic equilibrium. Private beachfront residential
developments were also rare because of the inaccessibility and absence
of infrastructure that characterized much of coastal America at the turn
of the twentieth century.

The value of coastal real estate and the extent of beachfront development rose dramatically over the course of the twentieth century. Federal,

state, and local governments, working through a variety of agencies and programs, expended unprecedented sums on hard infrastructure such as roads, bridges, and causeways that made formerly remote areas accessible, while on the shore, coastal engineers worked to stabilize inherently unruly shores—or at least give the appearance of stability. People flocked to coastal areas for the beach, and as coastal lands became an increasingly valuable commodity, so too did the beach.

The modern-day concept of the private beach and the fashioning of legal instruments for turning these commons into commodities originated in Connecticut. Beginning in the 1880s, wealthy families began building summer cottages along remote sections of shore in the state's eastern half. In 1885, the state legislature granted a charter to a group of families who owned cottages in Old Saybrook. The charter granted the Fenwick Association the power to levy its own taxes and enact zoning restrictions.[3] During the late nineteenth and early twentieth centuries, other small groups of families successfully petitioned the state legislature for charters to form what came to be known as private beach associations. Many of these early beach associations formed as an expeditious way of meeting the basic needs of summer homeowners in remote, undeveloped areas of the states lacking in basic infrastructure and services.[4]

James Jay Smith saw that beach association charters could also serve as an instrument for manufacturing and preserving exclusivity. In 1880, Smith started his own real estate investment and development firm and began buying properties in major US cities. By the 1890s, he had become a major dealer of real estate on Chicago's West Side, and by 1908, he was marketing waterfront lots along the Hudson River on New York City's Upper West Side. In 1909, Smith opened a real estate office in Old Saybrook and began buying up farmland and dense forested areas along the eastern half of the shore. With only a model for future development in hand, Smith petitioned the state legislature to grant special charters to imagined communities with evocative names such as Point O'Woods, White Sand Beach, Cornfield Point, and Grove Beach.[5]

With these charters, Smith's company marketed a prefabricated identity to prospective lot buyers, fortified by a host of deed and zoning restrictions and given concrete definition by a set of common, shared amenities and a form of private governance. In addition to giving its members the power to tax property owners and control land use, beach association charters also dictated the terms of membership and the election of officers, which included a board of governors, president, vice president, secretary, treasurer, and tax collector. Beach association charters also contained strict regulations on housing construction and land use, prohibitions against commercial activities and a whole host of other "nuisances," and exclusion of nonmembers from streets and land held in common by association members, including parks and, especially, the beach.[6]

Once chartered and populated, private beach associations had the power to perpetuate a certain identity and status. Like most homeowners' associations formed in the first half of the twentieth century, deed restrictions explicitly forbade the sale of lots to Black or Jewish people. Deeds in beach associations also contained provisions that owners of lots must be members of the affiliated beach club, with membership in the club determined by a board of governors. This helped to ensure that members could determine who could acquire property and become a part of a beach community and meant that most beach associations became closely associated with certain ethnic groups and religious denominations.[7]

Restricting who could buy property along Connecticut's shore extended to restrictions on who could access, traverse, swim, or relax along the beach. Private beach associations employed security guards to keep watch over the sections of the state's shoreline that fronted their developments and remove persons or groups that members deemed objectionable or undesirable. The African American novelist Ann Petry, who grew up in the coastal Connecticut town of Old Saybrook in the 1910s, was ordered to leave a private beach she was visiting for a picnic with an interracial Sunday School group. As the four-year-old Petry and other students sat in a circle, singing songs and playing in the sand, she recalled, a big "red-faced man" came marching toward her. "No n———rs allowed on this beach," he barked at Petry's Sunday School teacher. "It's writ in the rules," he added, warning the group to leave before he called the sheriff. Petry later described it as her "most humiliating Jim Crow experience."[8]

Keeping certain segments of the public off the beach became one of the main functions of beach associations' governing bodies, an abiding obsession of residents and officials alike, and one that was seen as inseparable from their desire to protect a community's "character" and enhance property values. Boards of governors "[worked] feverishly and with splendid dedication to preserve the pristine beauty of their private beaches and to frustrate invasions" of persons seeking access to the shore.[9]

Along with new restrictions on access to the beaches claimed by private beach associations, the spread of real estate development and the concentration of wealthy homeowners along the state's shoreline during these decades also gave rise to concerted campaigns to prevent the creation of public beaches. Beginning in the late 1920s in the town of Westport, along what came to be known as the state's Gold Coast, local officials, led by a band of wealthy homeowners, waged a protracted fight with the state to prevent the creation of a state park on the heavily wooded Sherwood Island located just offshore. When the state pressed ahead with plans, Westport officials hired a contractor to dredge a creek and flood the road connecting the state beach to the mainland. The move, one state official said, "will effectively prevent visitors from reaching the state property." Westport officials insisted that they were simply seeking to eliminate a mosquito breeding ground, but as another state official remarked, "The real object is to keep the people off state property." The people that local homeowners had in mind were those who could not afford to own a home in a Gold Coast town or summer cottage on the shore, for whom a state park with a beach was intended. Designating areas for the general public, as one Westport resident argued, "would be an invitation to the scum." Ultimately, the state succeeded in creating a state park on Sherwood Island, but it would be the last time it attempted to acquire coastal land for public use along the Gold Coast.[10]

The spread of private beaches influenced the ways that exclusive towns managed their own public beachfronts. Beginning in the 1920s, Greenwich and other Fairfield County towns passed laws banning nonresidents from using town beaches or placed other obstacles in their way. In 1930, Westport passed an ordinance that restricted parking privileges along the beach to residents only. The following year, it enacted an outright ban on nonresident use of town beaches on Saturdays, Sundays, and holidays.[11] Three years later, Darien adopted an ordinance limiting parking privileges at the town's beach to residents only. The ordinance, as its author explained, aimed to "effectively keep out all but residents and taxpayers of the town."[12] These same towns also enacted exclusionary zoning ordinances designed to prevent both the poor and people of color from living there. In some cases, towns worked in tandem with beach associations to restrict public access. In 1944, Greenwich purchased 147 acres on Greenwich Point (a.k.a. Tod's Point) for the purpose of creating a public beach. The Lucas Point Association, owners of a narrow piece of land (known as a driftway) that led to the point and which provided the only access from the mainland, acceded to Greenwich's plans to build a road to the point on the condition that the town "limit the use of the area to Greenwich residents." At a subsequent town meeting, Greenwich's board of selectmen passed an ordinance specifying that only "residents, taxpayers, lessees and their bona fide guests of the [t] own" were permitted to use the new beach.[13]

Resident-only beaches complemented and reinforced the exclusionary features of local housing markets. In Gold Coast towns, public officials, developers, realtors, and homeowners deployed a host of formal and informal measures to effectively keep both lower income families and people of color locked out of its housing markets. Homeowners and developers placed restrictions barring the sale of homes to Black people, Jewish people, and other disfavored ethnic groups in property deeds. Even after the US Supreme Court ruled such racial covenants unenforceable in the 1948 *Shelley v. Kraemer* decision, realtors continued to adhere to "gentleman's agreements" not to sell homes to racial and religious minorities, a practice dramatized in the eponymously titled 1947 film. Set in Darien, Connecticut, *Gentleman's Agreement* depicted the struggle of a white journalist posing as Jewish in finding a home in the town's "restrictive" neighborhoods.[14] In the decades that followed, anti-Semitism in Gold Coast housing markets decreased considerably. A 1964 study of Greenwich's housing market by the Anti-Defamation League of B'nai B'rith reported that "Jewish couples were treated as cordially and helpfully as . . . White non-Jewish couples" and "were shown substantially the same homes in the same neighborhood."[15]

African Americans, on the other hand, continued to suffer rampant discrimination in Fairfield County real estate markets. Realtors refused to show homes to Black couples, regardless of income or status. In 1953, Brooklyn Dodgers star Jackie Robinson and his wife Rachel were thwarted in their attempt to purchase land to build a home in

New Canaan, Connecticut, and were shunned by realtors in Greenwich. Only after area newspapers began reporting on the case of the man who broke the color barrier in baseball but who couldn't break into Fairfield County's housing market were the Robinsons able to purchase a home in the less exclusive town of Stamford. Still, by the mid-1960s, Fairfield County's Black population remained minuscule and mostly confined to isolated residential areas near downtown business districts.[16]

As a means of restricting access, preserving neighborhood exclusivity, and limiting negative publicity, large-lot zoning, Gold Coast towns discovered, proved far more effective. After the US Supreme Court authorized municipal zoning in its 1926 *Euclid v. Amber Realty Co.* decision, wealthier towns across metropolitan America established minimum lot size requirements and placed restrictions and outright prohibitions on multifamily units. In suburban Boston, historian Lily Geismer found, "wealthier towns during the interwar period . . . [adopted] rigid zoning and municipal planning laws to preserve both their physical characteristics and economic exclusivity."[17] For established towns seeking to ensure a high-income populace and preserve some of its rural characteristics in the midst of rampant suburbanization, minimum lot sizes held a special appeal. "By the simple design of large lot zoning," a pair of real estate economists wrote in 1961, "suburbanites believe that a municipality can achieve its developmental goals in a single stroke. The community will be beautiful, its taxes will be low, and 'undesirables' will be kept out."[18] In the decades following World War II, cities and towns in Fairfield County enacted ordinances limiting lot sizes to no less than two acres or, in many neighborhoods, four acres. During the 1950s, the average lot size in the suburban counties of the New York City metropolitan area (which included Fairfield County) doubled. By 1970, nearly one-third of Greenwich's 30,700 acres zoned for residential use required lots sizes of four or more acres. Eighty percent of New Canaan's housing market was zoned for two- or four-acre-minimum lots. Statewide, more than one-half of all land zoned for residential use during the 1960s had minimum lot requirements of one to two acres.[19] What critics called exclusionary zoning was most pronounced along the state's shoreline. A study by the fair housing advocacy organization Suburban Action Institute classified the zoning ordinances of every city and town along the state's shoreline, with the exception of Fairfield and Stamford and the port cities of New

London and Bridgeport, as moderately or severely restrictive.[20] In conjunction with residential exclusion, resident-only beach ordinances effectively amounted to a ban on both Black people and poor people from accessing much of the state's public shoreline.

Exclusionary housing markets drove up local real estate values and, with them, created large local tax bases relative to local population size. This supplied wealthy Gold Coast towns with ample tax revenues to fund local public schools, goods, and services, which, in turn, made those housing markets even more desirable and enhanced local real estate values even further still. It also had the effect of concentrating the state's poor and minority populations within cities and towns suffering from shrinking tax bases (from corporate relocation and deindustrialization, white flight, and urban renewal) coupled with rising public costs. In the face of declining local tax revenues and cuts in federal aid to cities beginning under the Nixon administration, cities like Hartford, New Haven, and Bridgeport were forced to relentlessly cut budgets, scale back services, and sell off public assets. Local parks and recreation departments and services were invariably the first on the chopping block.[21]

During these years, urban recreational programs and facilities struggled to meet the growing demands among its residents for access to safe and healthy outdoor recreation space. In its 1968 report on the causes of civil disorder in American cities, the Kerner Commission noted that the absence of safe, healthy, and attractive places for play ranked high among Black people's grievances, just behind police brutality, unemployment, and inadequate schools.[22] In Hartford, the absence of public parks and swimming pools in the area surrounding one of its largest public housing projects led children there to seek out fun and find relief from the summer heat by playing along the banks of the dangerous Park River, which snaked past their homes. Between 1942 and 1968, the river claimed the lives of seven children. Following each tragedy, parents stepped up demands for city officials to do something to address the crisis by erecting fences along the river and reinvesting in public recreation in low-income neighborhoods, but to no avail. In New Haven, high levels of pollution along the city's industrial shoreline forced the city to routinely close its public beaches.[23] Because of the ties that bound conditions in local real estate markets to municipal tax revenues, cities with large numbers of poor people and minorities lacked the resources needed to address these inadequacies. And because of the exclusionary practices of their wealthier neighbors, residents of the state's recreationally deprived urban centers had few options for relief outside of their cities.

As local municipalities clamped down on beach access, and as efforts by the general assembly to acquire and expand state beaches languished, the private development of Connecticut's shoreline proceeded at a rapid clip. By the late 1960s, the state was home to fifty-four beach associations and an estimated 184 private clubs and residential, nonstock corporations situated along the shore. Writing in 1961, Connecticut governor John Dempsey warned, "The time is not far off when the last remaining open area on Connecticut's shoreline is usurped for some private purpose." During these same years, publicly accessible sections of the state's shoreline disappeared. By the late 1960s, all but 7 of Connecticut's 253 miles of coast (and 72 miles of beach) were in private hands or effectively limited to residents of coastal towns.[24]

* * *

Wherever they were imposed, beach access restrictions generated controversy. Beginning in the late 1960s, local campaigns for open beaches and legal challenges of towns' and private homeowners' exclusionary measures proliferated across coastal America. Between 1970 and 1985, there were over 150 lawsuits filed in state and local courts in the United States that challenged beach access restrictions, as compared to approximately 10 during the previous seventy years.[25]

In 1969, Texas congressman Robert C. Eckhardt introduced a bill in the US Congress that would have established the public's right to access the nation's seashores. Modeled on the Open Beaches Act that he had helped to pass in the Texas state legislature in 1959, the National Open Beaches Act would have established, under federal law, the "free and unrestricted right to use" beaches; prohibited "fences, barriers, and other restraints on the use of the beaches by the public"; and empowered the Department of the Interior and the Justice Department to enforce these rights. The bill would have limited the power of local and state officials to rezone beachfronts for private development, restricted property owners from constructing fences or other obstructions across the beach to exclude the public, and provided states with federal funds to condemn land for easements to ensure public access. Despite garnering support in both the House of Representatives and the Senate, Eckhart's National Open Beaches Bill never made it out of committee, though watered down elements of it would be enfolded into the 1972 and 1975 Coastal Zone Management Acts.[26]

Local campaigns against coastal enclosure attracted a diverse range of allies, from environmentalists concerned about the effects of privatization and development on coastal ecologies to social justice activists seeking to tear down the legal and physical barriers that fed social division and inequality. Of these, few did more to call attention to the effects of beach privatization on the urban poor than the Hartford-based antipoverty activist Ned Coll. In 1964, the white Catholic recent college graduate abruptly quit his job working for an insurance company and founded Revitalization Corps, a self-styled domestic Peace Corps located in the heart of Hartford's Black ghetto. Revitalization Corps' programs ran the gamut from tutoring low-income children and helping young men and women find jobs to fighting slumlords and price-gouging merchants; from protesting austerity-minded politicians and indifferent bureaucrats, to providing free lunches to hungry children and boxes of clothes to needy parents.[27] After a group of mothers and caregivers asked for help in providing summer recreational options for area youth, Coll devised a plan to lead day trips down to the state's shoreline.

The first busloads of Black children from Hartford's North End began arriving at town beaches along the shore in the summer of 1971. There, they were greeted by locked gates and orders to leave. Shocked and incensed, Coll quickly turned his charitable endeavor into a protest against racist elitism and a campaign for open beaches. Each summer over the next decade, Revitalization Corps organized dozens of bus trips down to shoreline towns, where they demanded the right to access town beaches and took direct action against local exclusionary practices that, he claimed, violated their right to access public land. In response, shoreline towns like Old Lyme, Madison, and Greenwich adopted even more extreme mechanisms of exclusion, passing new ordinances and tightening existing ones, hiring more police and security, and mobilizing against any state or federal legislation perceived as threatening their local autonomy.[28]

These exclusionary practices had negative environmental effects. Coastal towns like Madison and Clinton ardently resisted adopting wastewater treatment facilities long after their towns' populations necessitated it. This was motivated in large measure by their determination to remain exempt from any state-mandated affordable, higher-density housing, by not having the capacity to provide it. As growing numbers of homes in shoreline towns were converted from seasonal to year-round occupancy, septic tanks malfunctioned with greater frequency, causing dangerously high levels of sewage to seep into Long Island Sound. Failing septic tanks caused once-robust shellfish beds to suffer sharp declines in productivity. During the 1974 summer season, dangerously high levels of pollution forced the state Health Department to ban the harvesting of clams, oysters, and mussels for the stretch of shore between Madison and Old Lyme.[29] By the late 1970s, officials in the state's Department of Environmental Protection (DEP) identified failing septic systems, in addition to other byproducts of overdevelopment, as one of the main causes of the water quality problems along much of the state's shoreline. In a report on the impacts of beach associations on coastal managements issued in 1976, the DEP identified a host of environmental problems resulting from uncoordinated real estate development among these hundreds of semiautonomous private governing bodies, whose "parochialism and narrow interests" ran counter to efforts to implement a "comprehensive [coastal] management program" that treated "the coastline as an integrated whole."[30]

While the waters of Long Island Sound suffered from pollution, the shoreline bore the scars from decades of rampant and uncoordinated overdevelopment. Jetties and groins constructed by individual property owners robbed neighboring areas of sand, disrupted the natural process of erosion and accretion, and instigated rapid changes to the shape and ecology of the shoreline, often with devastating consequences for coastal habitats. In towns such as East Haven, shoreline property owners waged relentless, if futile, battles against nature as they attempted to halt erosion and protect their homes from slipping into the sound. Fearing the sea and the state, vulnerable shoreline communities demanded, on the one hand, federal and state aid to protect their homes and, on the other hand, resisted the expansion of the state's environmental regulatory powers out of concern that it would force them to abandon their homes. By the mid-1970s, in East Haven's Momauguin neighborhood, severe erosion had turned the shoreline, which sported a sandy beach as late as the 1950s, into a jagged wall of boulders that grew higher and wider with each passing season. But residents nevertheless resisted calls to condemn and retreat, convinced that the dire warnings of coastal engineers masked ulterior motives. "Don't you realize the beachfront is gold-plated?" one homeowner commented to a reporter. "There is no more beachfront left on the East Coast." As soon as our property is condemned, he added, "the profiteers are going to come in, make a dollar, and walk out." Meanwhile, "With the money you get for your homes, you won't even be able to afford gasoline."[31]

In the early 1970s, the fate of coastal environments lay in the hands of state and local governments and private landowners. Aside from military installations, the only signs of federal power and authority in coastal America were the various beach replenishment and shoreline armoring projects being carried out by the Army Corps of Engineers, often at the behest of those governments.[32] The 1972 Coastal Zone Management Act (CZMA) established, for the first time, a set of federal standards for managing coastal land use and protecting coastal ecosystems. But it tried to do so by accommodating, rather than undercutting, local governmental power. An exemplar of President Richard Nixon's "New Federalism," which called for transferring greater power to states and localities to carry out federal policies and administer federal programs, the CZMA tried to strike a balance between federal standards and local administration. Instead of federal control of coastal areas, it offered states grants to establish and implement coastal management plans of their own, so long as they conformed to a set of federal guidelines.[33]

Connecticut would put the act's strengths and weaknesses to the test. When Republican state senator George Gunther introduced a bill to create a statewide Coastal Zone Management Council to formulate a coastal zone management agency, he was forced to respond to critics, who called it an attack on local autonomy. "I'm one of the strongest advocates of local autonomy," Gunther fired back. Shoreline public officials weren't convinced. "I feel that such a Council could and would usurp the powers of shorefront municipalities over our tidal waterfront land," Clinton first selectman Margery C. Scully warned Governor Meskill. "I also feel," she added, "that this Bill is just one more overt attempt to establish State zoning; and as with other Councils and Development Corps implies that local planning and zoning, recreation and conservation agencies, as well as the executive branch of the town, are incapable of making the right decision as to the highest and best use of our tidal shore and harbor front land. It takes common sense and vigilance on the part of local officials to guard against their growth and development . . . not State or Federal agencies or councils." Stamford's superintendent of Parks and Natural Resources Edward A. Connell, who was also the architect of the town's 1964 ordinance that effectively banned nonresidents from town beaches, emerged as a caustic opponent of statewide coastal management and led the charge to kill the bill and scuttle any efforts to impose outside authority on local shoreline practices. He implored Meskill to oppose the legislation, calling it "a gratuitous assertion by the self-appointed preservers of the American environment that the rugged citizens along Long Island Sound are no longer capable of planning their own communities." He issued a call to arms to the people of the shore to join in resisting "the efforts of the Superplanners to dominate the future land use of the State." "It is time for those with the salt spray in their nostrils to challenge the New Breed, fresh from Urban Planning courses in New York and Boston." What animated these fears? Madison's Vera Dallas made it plain: a statewide coastal management agency will, she predicted, "pave the way for opening beaches throughout the state."[34]

The coastal zone legislative committee identified political localism as the main threat to a healthy and sustainable shoreline. In its preliminary report to the state legislature shortly after the passage of the CZMA, it noted, "In too many cases, decisions for the establishment of policy and for changes in land use have been—and continues [sic] to be—the sole responsibility of individual municipalities. There is often no requirement and no incentive to consult other, even neighboring communities affected by proposals for development." Private beach associations, many of which functioned as separate government entities and jealously guarded the powers vested in them by the state legislature, exacerbated the problem of localism. "Some towns have as many as twelve to fifteen different bodies . . . which influence or have a voice in a particular decision such as development along the coastline." By one estimate, there were roughly five hundred individual administrative and regulatory agencies in Connecticut that were making independent decisions that affected the coast.[35]

In this political environment, lawmakers in Connecticut and neighboring states had long struggled to establish effective interstate regulatory bodies to manage and protect Long Island Sound. The first attempt was made in 1969, when New York congressmen Ogden R. Reid and Lester L. Wolff called for the formation of an intergovernmental agency known as the Long Island Sound Study Group that would conduct a comprehensive investigation of the sound's environmental, recreational, and industrial uses and conditions and produce a report outlining a set of policy recommendations for sharply reducing pollution and maximizing the sound's social, economic, and environmental assets. The proposal received strong support from Connecticut senator Abraham Ribicoff, who secured congressional appropriations to carry out the study. By the time it began its work in 1972, the New England River Basins Commission, the agency charged with coordinating the project, had a $13.5 million budget and a twenty-three-person committee appointed by the governors of Connecticut and New York and comprised of environmentalists, commercial and sport fishermen, businessmen, boating and swimming enthusiasts, civic leaders, and waterfront property owners.[36]

That same year, lawmakers in Hartford succeeded in passing a law regulating construction on inland wetlands. Passage of the bill reflected state residents' growing awareness and concern over environmental protection but also underscored their abiding commitment to local control. The law vested local commissions with the power to approve or deny building applications on wetlands and provided no oversight or auditing of local commissions' decisions by state authorities. While the law proved highly effective along the Gold Coast and in other regions with large concentrations of wealth, in other parts of the state, the allure of added tax revenue from development proved irresistible. A 1976 survey of fifteen towns found that, since the law had been enacted, five of the towns had approved every application submitted, while the other ten approved between 65 and 90 percent of all applications.[37]

The Long Island Sound Study Group also tried to accommodate the interests and prerogatives of shoreline governments. Following an outcry among shoreline communities at an initial draft released in January 1975, committee members cut entirely or watered down recommendations that the state work to acquire private property for public use and use the power of eminent domain to provide easements to the shore. The release of the final report later that summer garnered little fanfare and even less hope that it would lead to substantive action. Weighing over five pounds, and having taken five years and $3 million to complete, the final report's size and cost reflected its substance and significance. It found that the state suffered from "a serious shortage of general public lands along the shores of the Sound," and that, as a result, "large concentrations" of "low and moderate income families" were "recreationally deprived." Making the sound more accessible to the general public, the report's authors argued, was not simply a matter of fairness and equal opportunity, but also deeply tied to its environmental fate. "People who cannot see or reach the Sound to enjoy it cannot be expected to care about what happens to it or pay for its clean-up with their tax money. The Sound must be opened to provide more ways for more people, both city residents and suburbanites, to reach and enjoy it."[38]

But while the report was unequivocal in its call for greater public access, it avoided issuing any policy recommendations that might meet that objective. Instead, the report simply called on New York's and Connecticut's elected leaders to support the principle of public access. On the question that roiled local communities and consumed the energy of activists and civil liberties groups on both sides of the sound, the study's authors could only say, "Both state legislatures should declare it to be official state policy that the century-long trend of reduced public access be reversed." Even this general statement in support of public access alarmed residents of shoreline towns and beachfront homeowners.[39]

Throughout the late 1970s, the state struggled to create a coastal management agency, a precondition for the state joining the federal CZMA compact. At its core, the CZMA sought to bring an end to the uncoordinated, fragmented system that afforded shoreline towns and beachfront property owners total local control over coastal management. But this was precisely what allowed beachfront homeowners and shoreline towns to defy attempts to facilitate greater public access to the shore. Ultimately, legislators representing coastal districts agreed to pass the bill after its authors removed a passage that stated, as a general policy, the state's desire to "encourage public access to the waters of Long Island Sound and to encourage recreational opportunities within the coastal area." It would prove to be one of the weakest coastal management bills passed by any state.[40]

* * *

In 2001, Connecticut's Supreme Court ruled that towns could not bar the general public from accessing public beaches. But the state's shoreline remains, in many respects, as inaccessible as ever. Wealthy towns, especially those in proximity to poorer populations, have adopted a host of measures aimed at making it as difficult as possible for nonresidents to enter local public beaches, while at the same time ardently resisting attempts to expand affordable housing options. Beachfront homeowners continue to defy the law and the state's DEP, building fences and other obstructions across public lands in order to prevent the public from accessing the portions of the beach that front their homes. And the vigilance of private beach associations in patrolling their beachfronts remains unabated—as does their unpreparedness and incapacity to respond to the growing threats of a warming planet and its rising seas.[41]

NOTES

1 Ralph Yarborough, "Introduction of the National Open Beaches Act," in 115 Cong. Rec. 30335 (1969). See also Dennis W. Ducsik, *Shoreline for the Public: A Handbook of Social, Economic, and Legal Considerations Regarding Public Recreational Use of the Nation's Coastal Shoreline* (Cambridge, MA: MIT Press, 1974).

2 See Joseph L. Sax, "The Public Trust Doctrine in Natural Resource Law: Effective Judicial Intervention," *Michigan Law Review* 68, no. 3 (January 1970): 473–566; David J. Brower, William Dreyfoos, and Don Meserve, *Access to the Nation's Beaches: Legal and Planning Perspectives*, report prepared for Office of Sea Grant, National Oceanic and Atmospheric Administration, February 1978; Diana M. Whitelaw and Gerald Robert Visgilio, eds., *America's Changing Coasts: Private Rights and Public Trust* (Cheltenham, UK: Edward Elgar, 2005).

3 Incorporating the Fenwick Association, *Special Acts and Resolutions of the State of Connecticut, with an Appendix, Vol. 10, From 1885 to 1889 Inclusive*, House Joint Resolution no. 21, January 1885, 20.

4 Linda B. Krause, *Coastal Districts and Associations*, report prepared for the Coastal Area Management Program, DEP, December 1976.

5 "Chicago Real Estate," *Chicago Tribune*, April 6, 1890; "Developing Land in Old Saybrook," *Hartford Courant*, November 29, 1914; "James Jay Smith: Realty Developer Began Several Connecticut Shore Communities," *New York Times*, April 17, 1942.

6 "An Act Incorporating the Point O'Woods Association, Incorporated," HB605, Connecticut General Assembly, 1925, https://pointowoodsct.com/wp-content/uploads/2018/09/POW-amended-charter-6-3-18.pdf.

7 The James Jay Smith Company was the first real estate developer to implement the "club plan" in Connecticut; "Under this plan no lots will be sold to undesirable people, for it is a prerequisite for lot ownership that one must become a member

of the club." "New Development on Lake Hayward Attracts Buyers," *Hartford Courant*, June 8, 1930.

8 Ann Petry, "My Most Humiliating Jim Crow Experience," *Negro Digest* 48 (1946): 63–64.

9 "Friday Night Government," *Hartford Courant*, August 22, 1965.

10 "Westport Dredges Creek to Isolate State Park," *Hartford Courant*, August 4, 1929; W. H. Burr to W. O. Filley on Flood Plans, August 5, 1929, Connecticut State Parks and Forest Commission Correspondence, 1903–1963, Box 4, Folder: Correspondence, A. M. Turner, 1928–29, Connecticut State Archives; A. M. Turner to State Forestry Department, June 30, 1930, Connecticut State Parks and Forest Commission Correspondence, 1903–1963, Box 4, Folder: Januaray–August 1930, A. M. Turner, Connecticut State Archives. See also Kara M. Schlichting, "'They Shall Not Pass': Opposition to Public Leisure and State Park Planning in Connecticut and on Long Island," *Journal of Urban History* 41, no. 1 (2015): 116–42.

11 "Unneighborly Westport," *Hartford Courant*, June 9, 1931.

12 "Darien Moves to Keep Outsiders from Beach," *Hartford Courant*, May 13, 1933.

13 See Brenden P. Leydon v. Town of Greenwich et al. 257 Conn. 318, 777 A.2d 552 (2001).

14 *Gentleman's Agreement*, directed by Elia Kazan (Los Angeles: Twentieth Century Fox, 1947).

15 "Greenwich Tests Show Color Bias," *New York Times*, July 21, 1964.

16 Jason Sokol, *All Eyes Are Upon Us: Race and Politics from Boston to Brooklyn*, (New York: Basic, 2014), 64.

17 Lily Geismer, *Don't Blame Us: Suburban Liberals and the Transformation of the Democratic Party* (Princeton, NJ: Princeton University Press, 2014), 24.

18 James G. Coke and Charles S. Liebman, quoted in Michael N. Danielson, *The Politics of Exclusion* (New York: Columbia University Press, 1976), 59.

19 Danielson, *Politics of Exclusion*, 61.

20 Suburban Action Institute, *A Study of Zoning in Connecticut*, report prepared for the Connecticut Commission on Human Rights and Opportunities, New York, 1978, 83–85.

21 On the stratifying effects of fiscal fragmentation across mid-twentieth-century metropolitan America, see Werner Z. Hirsch, Phillip E. Vincent, Henry S. Terrell, Donald C. Shoup, and Arthur Rosett, *Fiscal Pressures on the Central City: The Impact of Commuters, Nonwhites, and Overlapping Governments* (New York: Praeger, 1971); Jon C. Teaford, *City and Suburb: The Political Fragmentation of Metropolitan America, 1850–1970* (Baltimore: Johns Hopkins University Press, 1979); Mark Schneider and John R. Logan, "Fiscal Implications of Class Segregation: Inequalities in the Distribution of Public Goods and Services in Suburban Municipalities," *Urban Affairs Quarterly* 17, no. 1 (1981): 23–36; Mark Schneider and John R. Logan, "Suburban Racial Segregation and Black Access to Local Public Resources," *Social Science Quarterly* 63, no. 4 (1982): 762–70; Andrew W. Kahrl, *The Black Tax: 150 Years of Theft, Exploitation, and Dispossession in America* (Chicago: University of Chicago Press, 2024), 73–127.

22 National Advisory Commission on Civil Disorders, *The Kerner Report* (Washington: US Government Printing Office, 1968), 8.

23 Andrew W. Kahrl, *Free the Beaches: The Story of Ned Coll and the Battle for America's Most Exclusive Shoreline* (New Haven: Yale University Press, 2018), 91–95.

24 Linda B. Krause, "Coastal Districts and Associations," report for the Connecticut Coastal Area Management Program, December 1976 (Hartford: Department of Environmental Protection, 1976); John Dempsey, "Gov. John Dempsey to Senate Cmte. on Interior and Insular Affairs," February 10, 1961, RG 005:034, Office of the Governor: John Dempsey (1961–1971), Box A-295, Connecticut State Archives; Kahrl, *Free the Beaches*, 27.

25 James Brooke, "When the Coast Isn't Clear to the Coast," *New York Times*, August 11, 1985; Gilbert L. Finnell, Jr., "Public Access to Coastal Public Property: Judicial Theories and the Taking Issue," *North Carolina Law Review* 67 (1989): 627–80; Marc R. Poirier, "Environmental Justice and the Beach Access Movements of the 1970s in Connecticut and New Jersey: Stories of Property and Civil Rights," *Connecticut Law Review* 28, no. 1996 (1996): 719–812.

26 Robert C. Eckhardt, "The Texas Open Beaches Act," in *The Beaches: Public Rights and Private Use* (Houston: Texas Law Institute of Coastal and Marine Resources, College of Law, University of Houston, 1972); Yarborough, "Introduction of the National Open Beaches Act"; Charles L. Black, Jr., "Constitutionality of the Eckhardt Open Beaches Bill," *Columbia Law Review* 74, no. 3 (April 1974): 439–47.

27 Kahrl, *Free the Beaches*.

28 Kahrl, *Free the Beaches*, 148–206.

29 Richard Matheny, "Case Study: Pollution of Hobbit Pond (Meadowville [Madison, CT])" (1973), Arthur Jack Viseltear Papers, Box 54, Folder 29, Yale University Library (Manuscripts and Archives); Connecticut Coastal Zone Management Committee, "Connecticut Coastal Zone Management Committee (Preliminary Draft Report)," August 14, 1972, Meskill Papers, Connecticut State Archives; "Septic Problems Force State Order Closing Facilities At Guilford Plaza," *New Haven Register*, July 18, 1974; "Madison-Old Lyme Off Limits For Shellfish," *New Haven Register*, July 12, 1974.

30 Krause, "Coastal Districts and Associations," 25; Michele Jacklin, "Engineers Pinpoint Septic Ills," *Hartford Courant*, October 19, 1978; "Zoning, Housing Issue Present Conflict," *Hartford Courant*, April 19, 1975; "Attorney to Give Talk oon Court Zoning Case," *Hartford Courant*, June 23, 1975.

31 Roger Hahn, "East Haven: Demolition in Momauguin," *New Haven Advocate*, May 30, 1979.

32 President's Council on Recreation and Natural Beauty, *From Sea to Shining Sea: A Report on the American Environment-Our Natural Heritage* (Washington, DC: US Government Printing Office, 1968), 175.

33 James A. Noone, "Resources Report/New Federal Program Seeks to Aid States in Control of Coastal-Area Exploitation," *National Journal*, December 9, 1972, 1889–98; David J. Brower, William Dreyfoos, and Don Meserve, *Access to the*

Nation's Beaches: Legal and Planning Perspectives (Raleigh: UNC Sea Grant, North Carolina State University, 1978), 40–43.

34 See Dan W. Lufkin to Governor Thomas Meskill, February 28, 1973, folder Coastal Management, box A-772, Office of the Governor: Thomas Meskill (1971–1975), Connecticut State Library (Hartford, CT); Margery C. Scully to Governor Thomas Meskill, February 16, 1973, ibid.; Edward A. Connell to Governor Thomas Meskill, March 17, 1973, ibid.; Edward A. Connell to Julius M. Wilensky (mayor, Stamford), March 17, 1973, ibid.; "Coast Panel Faces Opposition," *Hartford Courant*, April 1, 1973.

35 Connecticut Coastal Zone Management Committee, "Connecticut Coastal Zone Management Committee (Preliminary Draft Report)," 2–3; Krause, "Coastal Districts and Associations," 25.

36 Thomas J. McCormick to Governor Thomas Meskill, December 26, 1973, folder Long Island Sound, box A-774, Office of the Governor: Thomas Meskill (1971–1975), Connecticut State Library (Hartford, CT); New England River Basins Commission, *People and the Sound: A Plan for Long Island Sound*, report for the Long Island Sound Regional Study, 1975.

37 Connecticut General Assembly, Joint Legislative Review and Investigations Committee, *An Investigation of the Department of Environmental Protection* (Hartford: Connecticut General Assembly, December 21, 1976).

38 "Shore Owners Oppose Making Beaches Public," *New London Day*, January 22, 1975; New England River Basins Commission, *A Plan for Long Island Sound: Volume 1, Summary* (New Haven: New England River Basins Commission, July 1975), 22, 6.

39 New England River Basins Commission, *Plan for Long Island Sound*, 6; Chris Howard, "Ribicoff States It Would Be Tragic If Report on Sound 'Gathers Dust,'" *Greenwich Time*, July 29, 1975.

40 Kahrl, *Free the Beaches*, 207–30.

41 Kahrl, *Free the Beaches*, 263–304.

2

Cultural Repertoires and a Beach Dispute in Southern California

MATILDE CÓRDOBA AZCÁRATE

Local daily TV and radio news in San Diego, the second largest city in California, always includes the beach. The county has over seventy miles of beaches, divided in six different beach regions and over thirty-one county beaches distributed among its eighteen cities. Meteorological forecasts detail hourly surfing conditions and provide updates on recreational activities and beach cleanups. More often than not, and with alarming regularity especially after rains, reports also include beach advisories, warnings and closures caused by landslides and unrelenting erosion, collapsing cliffs, crumbling oceanfront property, sewage spills, and contaminated waters.

In early February 2022, an image caught my attention on the local news: a middle-aged, blond, sun-tanned, athletic woman dressed in sports apparel with sunglasses was talking to a local TV reporter, visibly agitated. She was atop of a bluff overseeing one of San Diego's finest beaches in Del Mar right next to the rail tracks that travel along the coastline. She was part of a small group gathering and she was holding a medium size banner with a handwritten message that read "Don't Fence Us Out." She was oscillating the banner side to side with intent, as if wanting the message to urgently trespass the screen.

As I watched, I thought the moment seemed odd, somewhat counterintuitive. Why would this white, presumably well-off, woman be saying "no to fences"? Were there plans to build a fence on the beach? As urban anthropologists and geographers have shown, fences, gates and walls have historically kept wealth contained, landscapes and resources enclosed, for the benefit of the privileged and labor invisible.[1] In the Southern California I've come to know as a resident myself in these past nine years, fences are the ultimate material infrastructure designed to

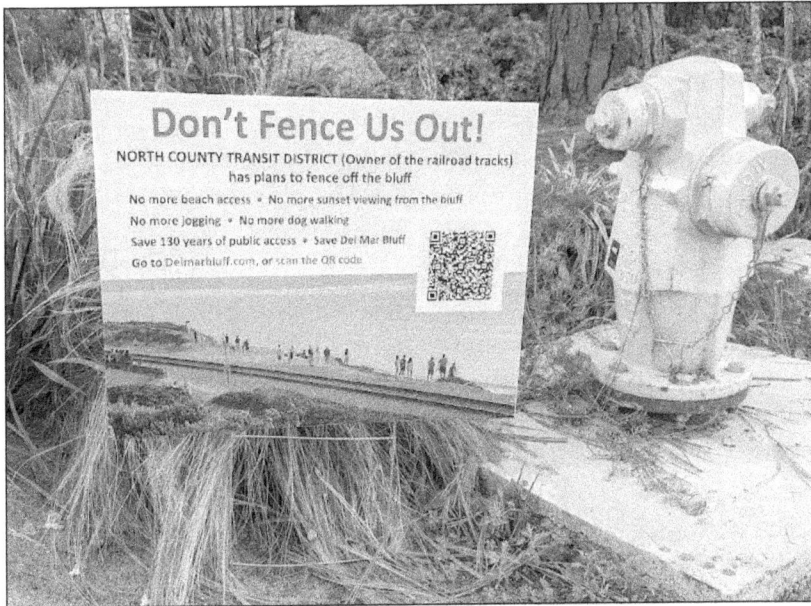

Figure 2.1. Yard sign in Del Mar opposing the fence. Source: *Times of San Diego*, Chris Jennewein, January 2022.

signal who belongs and who is just a passerby. Why would someone who looked like she belonged to this privileged landscape, to this wealthy beach, in the normative stereotypical way, be *opposing* a fence?

The counterintuitiveness of this moment left me intrigued. By doing some digging, I learned that the woman was a Del Mar resident, born and raised, and part of a collective named "Del Mar Fence Fighters," a wealthy and well-organized group mostly made up of Del Mar residents who opposed a 2018 regional proposal by the North County Transit District (NCTD) to fence part of the railroad tracks that, when crossed illegally, as was customarily done by some Del Mar neighbors, granted local residents direct access to the beach. Now, this made more sense. The so-called Del Mar Fence Fighters were claiming what they perceived to be, and had de facto experienced as, their legitimate right to the beach (figure 2.1). Their fight against the fencing proposal came from privilege and was intended to maintain a practice that, despite legal and safety concerns, the fence would interrupt. They had fighting on their side, I later learned, other major nonprofit and volunteer associations in the city, as well as institutions and activist networks devoted to environmental conservation and social justice, such as the California Coastal Commission, the San Diego Association of Governments, and the Surfrider Foundation. Why and how had they come together?

Since I came across it, the Del Mar dispute has nurtured a larger research project I have been working on. This project looks at refusals to state-planned rail infrastructures perceived by local populations to encroach the use of public spaces, including beaches. Its main focus is the mega-infrastructure of Train Maya, a novel tourist train that will connect the main tourist spots along Southern Mexico that is being fiercely contested by scientists, international agencies, residents, and Indigenous constituencies for its environmental and sociocultural costs.[2] Refusals to Train Maya are articulated as demands from the dispossessed against state intervention. In Del Mar, it seemed quite the contrary. Refusals to state-led infrastructure development came from the privileged.

The Del Mar dispute over access to the beach and over rail tracks brings to the fore the beach as a political and politicized terrain where multiple constituencies with seemingly disparate interests come together. But, when, how, and why are beaches mobilized to make claims for rights—rights of access, of belonging, of mobility justice, of environmental sustainability? Are there other grammars beyond "rights" to make sense of beach politics? And if so, which ones? What form and shape can infrastructural refusals to state-led interventions *at* and *in* the beach take? In this chapter, I build on the Del Mar beach dispute as a smaller case study in a preliminary effort to develop some of the analytical tools needed to make sense of refusals to infrastructural projects that center around the beach. I use the concept of *cultural repertoires*, or the supply of skills and devices used by different social actors to enunciate and communicate competing views and demands on the beach as a physical, symbolic, and affective social space.

I build on the sociological literature on the role of repertoires in sociocultural analysis, particularly on the pioneer work of Ann Swidler on culture as a toolkit and Michelle Lamont and Laurent Thevenot's broader comparative cultural analysis of repertoires of evaluation in France and the United States.[3] Repertoires, as Lamont and Thevenot put it, "are elementary grammars that can be available across situations and that pre-exist individuals, although they are transformed and made salient by individuals."[4] These grammars enable researchers to observe and point at "available cultural positions that are not centered around political institutions" necessarily.[5]

In what follows, I discuss three different yet related cultural repertoires at play in Del Mar's fencing dispute that might be of use to understand other beach disputes as well. I have been able to identify: (1) *repertoires of cultural fixation*, which are intended to secure the beach of the status quo: a white, enclosed, residential space and globalized tourist attraction; (2) *scientific repertoires*, which monitor and predict cliff activity according to Western-centric tools such as metrics and projections and aimed at climate mitigation; and (3) *fugitive repertoires*, which emerge broadly in Indigenous-led relational enunciations of water stewardship. These repertoires are neither mutually exclusive nor are they used by a single set of actors, be it scientists, activists, or residents. In the Del Mar case study, they operate in nested ways, often through fuzzy and dynamic boundaries.

As I have been able to trace, repertoires of cultural fixation and scientific and fugitive repertoires are mobilized by different actors within and outside the physical space of the beach in order to make both individual and collective claims about land ownership, access to public spaces, environmental education, knowledge production, and the future. Sometimes, these three repertoires work together to shift ingrained discourses, for example, that of the scientist-activist divide. Sometimes, they serve to make sense of scientists-as-state-actors or of Indigenous practices that contribute to evade the status quo beach in favor of relational and nonextractive understandings of space and place, sand, and water. Out of these three repertoires, the first two, repertoires of cultural fixation and scientific repertoires, enable the identification of grammars of already available cultural positions, while the third one emerges as a grammar still to be fully formulated. The three of them, as I will show, are also strongly mediated by new technologies, and they are mobilized by individuals and collectives that use the virtual arena to amplify a hold on their visions and practices over space and the future: uneven practices of beach access, scientifically driven climate mitigation plans, or the promise of emancipation from colonial histories of land and ocean dispossession, respectively. All of them share the characteristic of explicitly wanting to act on political decision-making and are oriented toward the future, a future that current weather events and associated disruptions announce as dire. For this reason, I understand cultural repertoires as socio-material and political technologies of future-making.

Tourism, Privilege, Dispossession, Beach Erosion and Water Stewardship

Mike Davis, "California's prophet of doom," described this region, and San Diego in particular, as the perfect mix of extreme wealth and brutal dispossession; of tourist glamour and inhumane poverty; of mesmerizing nature and extreme anthropogenic ecological destruction and violence.[6]

As a tourist destination, the coast of Southern California is, as Mimi Sheller and John Urry would put it, a paradigmatic "place to play and a place in play" in the global tourism market—that is, a place "made and remade by the mobilities and performances of tourists and workers, images and heritage, the latest fashions and the newest diseases."[7] This is a tourist reality enabled by the erasure of Indigenous forms of life through violent colonial processes of resource and labor extraction, land dispossession, and genocide.[8] Inscribed in the landscape in the form of missions, casinos, and overlooked burial sites, these erasures have been furthered under capitalism's uneven development, selective heritage preservation processes, and the militarization of the coastline. They are moreover clearly informing both contemporary social interactions and the relations that different social groups have with spaces such as its beaches. Contemporary land use regulations, labor, city politics, and urban planning in and around San Diego County, for example, are mostly guided by tourism goals and, while centering around its beaches as the major attraction, they transcend its physical space to inform housing decisions, transportation, provision of public services, and memory.

Southern California's past of erasures and simultaneous present-day tourist *and* military realities have enabled the production, distribution, and consumption of some of San Diego's northern beaches (those farther away from Tijuana, Mexico, such as Del Mar) as materially and symbolically privileged, globalized, and largely white spaces. In my analysis of Del Mar fencing opposition in this chapter, this particular configuration of the beach matters. It matters because, as previous research has shown, beaches are instruments for social justice and democratic transformation. Waleed Hazbun's and James Freeman's research on the Middle East and Rio de Janeiro, respectively, shows that beaches are sites "of an unequal, often confrontational politics of class whereby the legitimacy of the social order is challenged, renegotiated, and ultimately reproduced."[9] Beaches in Southern California fit this description. They are, furthermore, spaces where rights are negotiated, as Ajantha Subramanian illustrates for India, and where affects unfold, as Bianca Williams describes for Jamaica.[10] In California, as Christina Dunbar-Hester has recently shown, the toxicity of logistic infrastructures threatens ecological life, subsuming it to the principles of capitalist consumption in the Port of Los Angeles and Long Beach particularly.[11] A bit south of these places, collapsing cliffs over tourist beaches show reminiscences of what the future holds in a strikingly similar fashion (figure 2.2). Beach erosion and collapsing cliffs are at the core of Southern California's failing transportation and housing infrastructures along the coast connecting San Diego to Los Angeles and beyond. For this reason, cliff monitoring has been the focus of increasing scientific interest in the last few years, especially since the region has experienced a multiplication of devastating climatic events such as severe droughts and wildfires, historic rainfalls, tropical storms, and unusually large king tides. These meteorological phenomena have visibly transformed beaches and their uses, altering or fully closing them for human use. In the last two years, for example, road and walkway access to most beaches has become impractical, fully covered with rocks, cobbles, and sand; unstable cliff banners are ubiquitous; and water advisories warn about exceeding safe levels of bacteria almost daily, everywhere, making beach closures no longer an exceptional event.

Climatic events, added to predatory human patterns of coastal occupation and infrastructure development, have prompted recent deaths among beach goers in San Diego due to massive, sudden cliff collapses. The most controversial collapse happened in Grandview Beach, Encinitas, a neighboring community to Del Mar, killing three members of the same family in 2019. The recurrence and spectacle of these collapses have granted beaches in Southern California even more space in regional, national, and international news and public debates, where Southern California emerges as an iconic site of anthropogenic ecological neglect and threatened global tourist destination. These spectacular events and

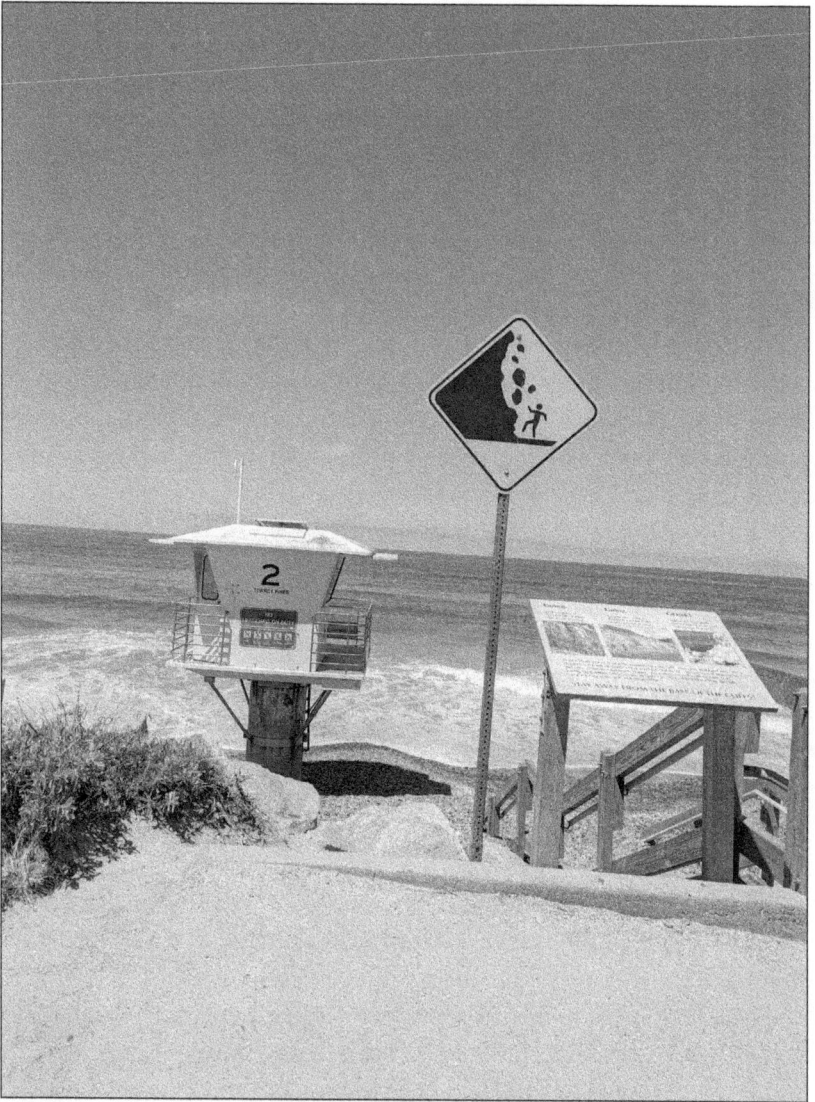

Figure 2.2. Unstable cliff post, Torrey Pines State Park, California, March 2023.
Photo by Matilde Córdoba Azcárate.

their spectacularized narration in media outlets contribute to the fixation of the beach as an enclosed, racially marked space for leisure and conservation, where agency is primordially given and expected from white wealthy bodies and Western science—or what is the same, from those with access to experience its beauty and now alarmingly robbed of it, and from those with the knowledge and expertise to monitor its behavior and hence capable of preventing such disasters.

These stories and media representations, however, silence and erase Black and Indigenous conceptualizations and experiences of the beach otherwise. They displace the attention from other existing coalitions of science, activism, and citizen alliances that have been calling attention to the beach beyond this particular visualization. But before getting to them, lets situate the dispute.

The Dispute: "Don't Fence Us Out"

The 2018 North County Transit District (NCTD) fencing proposal in Del Mar is part of a larger Railway Safety Enhancement Project intended to increase public safety and avoid trespassing, or, as the proposal put it, to "help minimize deaths, injuries, and fatalities that occur by trespassing on and/or illegally crossing railroad tracks" to access the beach (figure 2.3).[12] This is a practice that the NCTD identified as happening regularly since the 1990s in the cities of Del Mar, Encinitas, and Oceanside, three extremely wealthy and predominantly white cities on the Southern California coast. The NCTD fencing proposal found justification as part of larger ongoing federal improvements along the California coast that also articulated residents' and tourists' safety as major reasons for infrastructural intervention. As the NCTD explained, for the Del Mar fencing proposal, the right-of-way fencing was supported by a 2020 Trespassing Risk Mitigation Study, an "assessment of risk and high-level engineering feasibility" that had identified this beach city, and the neighboring Encinitas, as one of "the areas with the most train strike and trespassing incidents in San Diego County."[13]

Since its formulation, the proposal was fiercely opposed by residents in Del Mar and the California Coastal Commission, which is in charge of planning and regulating the use of land and water along the coast and is committed, as they put it, "to protecting and enhancing California's coast

Figure 2.3. Rail crossing safety warnings, Cardiff by the Sea, California, April 2023. Photo by Matilde Córdoba Azcárate.

and ocean for present and future generations . . . through careful planning and regulation of environmentally-sustainable development, rigorous use of science, strong public participation, education, and effective intergovernmental coordination."[14] The fencing proposal, they argued, lacked environmental assessments, and the commission had not been made aware of it until it was presented as something already decided.

In the fencing plan, NCTD elaborated that the cities involved would have the opportunity to decide what type of fences would best fit their needs: either engineered standard fences (galvanized chain-link fencing) or alternative fence options (redwood post and cable fencing; vinyl coated chain-link fencing). For the City of Del Mar, however, there was no option to be made. NCTD signaled a 1.4-mile-long, six-foot-high black vinyl chain-link, like the one at the nearby popular Powerhouse Park—a historical public park available for events, with pay parking, playground, restrooms and showers, and, importantly, easy pedestrian beach access. The Del Mar fence was also the highest fence of those proposed (with none of the others higher than four feet). These elements, combined with the lack of environmental and citizen consultation, generated an immediate civic backlash.

Soon after the proposal was presented to the City of Del Mar in 2018, city residents became organized. A new, relatively small yet politically powerful collective named Del Mar Fence Fighters was created, as well as different online platforms.[15] These included a 2020 Change.org petition which gathered over six thousand signatures, a new platform named delmarbluff.com that invited residents to join opposition, and the Friends of Del Mar Bluffs nonprofit. On these platforms, residents and council members representing the City of Del Mar and the California Coastal Commission opposed the fence building mostly with narratives of libertarian freedom and environmental conservation.

For organized residents, the proposal, and the fence itself, was a direct insult to their individual freedoms and cultural community practices. The fence, they claimed, "would obstruct beach access, denigrate pristine ocean views, and potentially lower property values and discourage tourism."[16] Furthermore, the fence aimed to put a stop to what some neighbors described as "a century-old practice"—that of *illegally* hopping railroad tracks to reach the beach. Fences, they argued, were unnecessary because "people have learned to look for trains and safely cross the tracks or walk the trains beside the railroad."[17] Those *belonging* to that beach thus knew how to hop the rails, and accidents and deaths happened among newcomers, visitors, those without the know-how.

Backed by the California Coastal Commission, residents also opposed the fence using environmental arguments, most notably, those of coastal erosion. The infrastructural interventions needed to secure the fence, they said, posed a threat to already unstable cliffs, and these had already suffered enough anthropogenic interventions and were at the brink of total collapse. Drilling to place the fence, they pointed out, would have nefarious effects in already eroded cliffs and thus contribute to more beach closures and further disruptions to cherished practices such as sunset viewing, surfing, jogging, and walking dogs.

In 2020, the conflict between Del Mar residents and NCTD reached the Surface Transportation Board. Described as a "hot potato" decision in local media, the safety fence prompted two civil lawsuits in the San Diego County Superior Court, one in 2021 and one in 2022.[18] As a direct result of these legal and social mobilizations, the NCTD proposed a scaled-down, four-foot-tall design to offer, as they said, a more "community-sensitive" approach to the initial fencing plan. Yet, as local media put it, "most of the community wasn't swayed" and mobilizations and litigations continue as of the writing of this chapter.[19]

Railroad fencing opposition in the beach city of De Mar is the result of a history of privilege and white settler occupation of the coastline in Southern California that culture, tourism, and science have helped to secure in its libertarian configurations. This citizen-driven opposition to rail fencing, to put it in other words, is ultimately motivated by a plight to individual freedom to continue with informal (and illegal) practices of access to the beach mostly performed by wealthy white residents. Their claims to the beach under such configuration became urgent to readdress in contexts of climate emergency and postpandemic tourist reform, when coastal rail transportation and access to the public beach became unquestionably threatened. But what can this dispute teach us about beach politics more generally? I turn to the notion of cultural repertoires to think about this and to describe the coalitions of social actors that gained (direct and indirect) visibility in its wake.

Repertoires of Cultural Fixation: The Status Quo Beach

Neoliberal orderings of public space are a constant in the history of the United States and beyond, but when it relates to beaches, their imagination as public open-access spaces, as "the ultimate fantasy space," helps to occlude the experiences of exclusion that are often times not only experienced on an everyday basis, but importantly, legally supported.[20] This is the backdrop against which organized neighbors in Del Mar oppose railroad fencing through public demonstrations and online activism. The cultural repertoires they engage with participate in processes of neoliberal cultural fixation that aim to secure the beach as a white and wealthy enclave for their recreational use.

The beach Del Mar Fence Fighters refer to is the California status quo beach, the one depicted by tourism marketing and cultural industries as that white, leisure and surfing space above all else. This is a hegemonic cultural representation, which has dominated and fixed how California beaches exist in local and global imaginaries. Importantly, this hegemonic cultural representation is only possible through the erasure of difference, both present and past. In appealing to and using this hegemonic cultural representation of the beach as part of their toolkit to oppose the fence, organized neighbors in Del Mar secure their monopoly over this public space, which demands the exclusion of others, including, in this case, the state. This is why this representation is not neutral, and what they are doing in claiming "Don't fence us out" is not neutral either. The point is, then, to ask how repertoires of cultural fixation are strategically used and selected to articulate a protest without altering the status quo, thus using the preservation of the beach as a way of preserving privilege, meaning allowing access just as it is.

Media reports on fence fighters' demonstrations in Del Mar show how opponents to the fence mobilize hegemonic cultural referents about the beach to their advantage. In their public protests, residents regularly repeated "Fences don't boost tourism right?" and "Don't fence us out."[21] According to them, the fence disrupts both tourists' rights for the enjoyment of the beach and scenic views as well as historical practices of pedestrian access to the beach. Some residents argue that beaches along the Pacific coast of Southern California are "global icons," celebrated in North American popular culture, and that the fence would irremediably threaten those representations. Other residents directly appeal to North American popular songs, most notably the Beach Boys' "Surfing U.S.A," which has a direct reference to Del Mar. Building on this popular culture referent, residents describe their learned practices of accessing the beach by hopping railroad tracks as "part of the culture of the town." In doing so, they also appeal to the pioneering nature of their city, "a family-oriented community" as official documents read, founded as a resort town by a rancher and a railroad official in the early 1880s.[22]

Along with allusions to learned practices, a history of pioneer settlement, and popular culture, fence fighters also used a shared and emotionally charged narrative about the beach as a healing space during the COVID-19 pandemic in order to mobilize other residents in opposition to the plan. This narrative is emphasized, for example, in their 2020 Change.org campaign (with over six thousand signatures), which read, "The pandemic has been tough on everyone, but the beach has been an important source of relief during these stressful times. Limiting access is unacceptable."[23]

Cultural allusions such as these are problematic since they equate the space of the beach with that depicted by and for a mostly middle-upper-class and white North American public. In appealing to settler history, popular culture, tradition, and customary rights, Del Mar Fence Fighters avoid discussions of the illegality of the crossing practice itself, as if leveraging this popular, global, and libertarian representation of California beaches was enough recourse to legitimize their practices all along.

In the complaint filed against the fencing proposal by Del Mar residents, cultural arguments came to the fore when organized residents wrote that the fencing "planned for the Upper Bluff and Lower Bluff would be damaging to the environment, including by permanently altering the bluffs, accelerating erosion, and creating artificial barriers with a prison-like industrial appearance."[24] This allusion to science and prison aesthetics is neither neutral nor anecdotal. It appeared often in residents' interviews with media journalists, where residents claimed that the fencing proposal wanted "to decapitate the bluff . . . and turn Del Mar into an industrial zone," "killing views" and resulting in "imprisonment" of the community.[25] By mobilizing the opposite representation of California's beaches depicted in songs like "Surfin' U.S.A." as a fundamentally white space where freedom is exercised, these discourses intentionally mark industrial California and the prison industrial complex as nonwhite and toxic spaces at the same time that they erase Black and Indigenous bodies from the beach. Their right to the beach as they know it implies the absence of these constituents and their rights—their rights to access the beach, to enjoy it, and to imagine it otherwise.

Repertoires of cultural fixation might be present in other cases where access to the beach is disputed. Paying attention to them and asking what it is exactly that they aim to fix might enable us to identify social actors closer to the status quo as well as to spell out the arguments they use with respect to who has the right to the beach and how that right has been exercised in the past and in the present. It also might enable us, as it does in this case, to pinpoint what is the type of beach they envisage for the future and who is included and excluded from it and how.

Scientific Repertoires: Ensemble Predictions and Beach Memory

In the Del Mar fencing dispute, there are also related repertoires that read the beach through the lenses of science. These repertoires that I call scientific mostly aim to estimate shoreline behavior based on averages (of waves' behavior for example) that build up what natural scientists have called *beach memory*.[26]

Scientifically driven arguments are used by those opposing the fencing project in Del Mar to point to already vulnerable coastal bluffs. These arguments mostly rely on metrics and global statistical indicators to predict coastline behaviors and postulate social solutions in a direct effort to mitigate further environmental damage of coastal landscapes. Examples of scientific repertoires at play are, for example, real-time ensemble predictions and projections on cliff erosion done by prestigious research universities in the region.

Ensemble predictions "are numerical weather prediction (NWP) systems that allow us to estimate the uncertainty in a weather forecast as well as the most likely outcome."[27] They are broadly used, and not without contestation, within scientific frameworks of climate adaptation and climate mitigation, which some scholars have already shown we should understand as cultural artifacts themselves, since technical knowledge and vulnerability are always embedded in webs of settler colonialism in the United States and beyond.[28] Ensemble predictions imprint in the landscape old, colonial ideas of control, mastery, and domination over nature that, as Amitav Gosh put it, seem to have "been incorporated into the very foundations of middle-class patterns of living across the globe," particularly in coastal cities.[29] This incorporation is evident in the arguments of those against the fencing proposal, including the California Coastal Commission. Even before recent cliff collapses and tourists' and residents' deaths at the beach, coastal officials, residents, and businesses in Del Mar had turned to experts for advice on the cliffs' projected behavior in order to plan for work on roads, railways, housing, critical public infrastructure, and private dealings. In a cease-and-desist letter demanding the end of the NCTD's attempts to proceed with the fence, the California Coastal Commission highlighted the lack of a coastal development permit and the need for residents, environmental experts, and the NCTD to work together toward Del Mar bluff stability. The let-

ter was ignored, which resulted in a major lawsuit where the state commission accused NCTD of noncompliance with the California Coastal
Act and state and local development policies.

Since 2018, further state resources have been devoted to coastal cliff
forecasts in the areas of Del Mar, Encinitas, and Oceanside, where the
fencing projects are taking place and where most of the scientific research is condensed. Scientific research on coastal erosion has become
a leading factor in municipal political campaigns, and data produced
by research universities is used by residents, environmental activists,
and local political parties of all sides to make decisions and establish priorities in beach communities all along the coast. In San Diego,
science-driven knowledge on cliff behavior, for example, has been used
to offer politically driven calculations of risks and probabilities of collapsing bluffs. Although these forms of knowledge are contested within
research teams and scientific paradigms, once science is mobilized for
public opinion, monitored cliff behavior becomes a "fact" frequently
and homogenously highlighted by Del Mar residents when opposing the
NCTD's fencing proposal. The City of Del Mar, they say, has historically
engaged in emergency bluff repair projects in coordination with experts
and without any state or federal support. Hence, it has helped NCTD's
safety procedures all along, yet not in this particular project because
they perceive that their customary cultural practices are under attack. By
recourse to natural science, Del Mar neighbors opposing the fence make
their claims with a legitimate tool that protects their interests without
threatening the status quo California beach—the one that has historically and purposefully erased other bodies, other stories.

Ensemble predictions on cliff activity in Del Mar become mobilized
as what Stephanie Kane has defined as "political engineering" exercises
in which the technical gives way to political decision-making that is
based in neither geophysical nor socially unique analysis; beach memory is not just recorded by science, measured and anticipated through
scientific techniques.[30] In this case, scientific repertoires amplify liberal
views about space and place and, as Katherine McKittrick would say,
they must be nurtured by other stories because they are partial and dis-
proportionally keep privileging racialized and normative ideas of space
and place,[31] in this case, the beach as a leisure and conservation space
for the white and wealthy per North American parameters.

Indigenous Repertoires: "The Water Is Flowing, the Water Is Mixed"

While the Del Mar beach access dispute was developing by gaining traction in local and regional news, unprecedented extreme weather events severely affected the coast of California. Droughts, massive fires, landslides, and even more frequent cliff collapses in northern San Diego beaches including Del Mar interrupted train service along the coast for months, disrupting commuters, homeowners, residents, and visitors alike. Simultaneously, alternative narratives about the beach proliferated. They related a beach that had nothing to do with libertarian visions of space or with land property claims except for the fact that they had been suppressed by them, erased or buried. These narratives sprung from Indigenous repertoires and worked in a fugitive manner, aiming to evade and escape their silencing and suppression by the California status quo beach for tourism and of science.

Maya Berry and collaborators elaborate on ideas of fugitivity as an embodied feminist ethos and decolonial path to escape critical anthropological methodologies that remain complicit with race, gender, and class status quo.[32] Raymond Craib describes "fugitive landscapes" as those "improvised, indeterminate, and administratively intangible" areas that the state aims to fix in vain through cartographic representations.[33] I build on these complementary uses of fugitiveness to describe these other cultural repertoires that emerged in the light of Del Mar fencing opposition, even when not directly enunciated in relation to it.

These fugitive repertoires are often Kumeyaay-based, and they come to the fore in this larger process of absence-presence when dealing with the beach. These repertoires make visible and narrate the beach through Indigenous situated practices of place-making in shared contexts of land dispossession and historic genocide. Fugitive repertoires not only point at already existing cultural milieus but, importantly in this case, they curate the imagination of beaches otherwise—for example, as common spaces of rootedness and future-making that exceed current frameworks of resource extraction, uneven capital accumulation, and dispossession.

In Southern California, Indigenous water stewardship offers alternative conceptualizations away from capitalist's extractive uses at the core of exclusivist oceanfront property development that have fueled beach erosion processes and toxic sewage spills into ocean waters. Considering that most cities, including Del Mar, are built upon occupied Native land, media attention to beaches as Indigenous spaces is still small in San Diego and completely absent in this fencing dispute. Research universities, Indigenous constituents, and social movements, however, had been already working hand in hand for some time when the dispute erupted. They came together to educate the public about the historical relationship of Indigenous people with the ocean. Education programs included events at different beaches in San Diego County, including La Jolla, next to Del Mar, as privileged pedagogical spaces not only to talk about Indigenous relational and holistic understandings of nature but also, and importantly, to build communities and awareness that the beach, the same one Del Mar Fence Fighters fight for, is also, and foremost, theirs.

Two examples of Indigenous repertoires that gained media attention at the same time as the Del Mar fencing dispute were Kumeyaay initiatives to educate the public on Indigenous custodianship with respect to the ocean and its resources through hands-on activities and storytelling practices, such as Whose Coast Are You Surfing in San Diego? and Tribal Water Stories of Coastal Southern California, respectively. These two initiatives have built alternative repertoires that are now at hand for residents, scientists, visitors, and other collectives in their individual and collective engagement with the beach beyond normative representations–part of the toolkit to read and practice the beach and hence disputes like the one at hand.

Whose Coast Are You Surfing in San Diego? is part of the InterTribal Youth's Native Like Water fellowship program, which educates Indigenous youth on science, conservation, and culture in the San Diego area. It is a collaboration between a history professor at Kumeyaay Community College and Native Like Water and InterTribal Youth founder. The initiative focuses on both youth and adult cultural exchange between ocean and fresh water land environments and aims to "indigenize education" through a hands-on approach.[34] Native Like Water organizes activities at the beach and through a collaboration with the Canadian nonprofit Native Land Digital, which provides access to an interactive app that navigates Indigenous territories, languages, and treaties educating the public about whose coast are they surfing and beyond.[35] "As surfers," Native Like Water states in their public documents, "we love to tell

Figure 2.4. Tule Canoe Building, San Diego, 2021. Photo by Matilde Córdoba Azcárate.

each other to respect locals . . . but the fact is that all non-natives in San Diego are living (and surfing) along a coastline that was taken from the Kumeyaay people."[36] Kumeyaay origin stories, they say, suggest a beginning in the waves, and it is paramount to educate the public about the relationship between Native people and the ocean. Since 2017, they have organized activities, such as tule canoe building, as well as prayers and festivals to oppose the further construction of the US-Mexico border wall just a few miles south from Del Mar (figure 2.4).

Tribal Water Stories of Coastal Southern California is a compilation of Kumeyaay origin stories that relate to water.[37] This is a relatively small but comprehensive and open-access repository of songs, stories, and myths of Native American Tribes' relation to water that was created by the San Diego Integrated Regional Water Management Program, an interdisciplinary effort of organizations and stakeholders invested in ameliorating water politics in the region. The compilation is a result of the Tribal Water Summit of 2009, with papers published in 2010. The summit as well as the compilation of stories have a political goal in mind: to escape the normative practice of leaving Indigenous people out of water planning and policy by restoring their water bonds through inclusive policy making that envisages Indigenous constituencies as water funding beneficiaries. The project, which continues to this day, also aims to shift current perceptions of Indigenous peoples as being only bounded to land in Southern California and to educate the public about their significant relationships with water, including the ocean. Stories and illustrations in the repository start and end with the ocean, reminding readers that water and the ocean have been at the forefront of Kumeyaay stewardship practices and that Indigenous peoples also belong in the space of the beach. This is a space experienced, felt, and conceptualized in a relational manner, or, as the compilation puts it, in honor of Kumeyaay song writer Gloria Castañeda Silva in her depiction of water, as a "flowing" and "mixed" space.[38]

In these Indigenous conceptualizations, ocean water is not described as an economic resource nor is the beach perceived as just a leisure space, an economic resource, or a space for expert conservation, but as a socio-atmospheric element that is in relation to others. This relational understanding of nature clashes with the Del Mar Fence Fighters' property driven claims and, paradoxically, it has gained traction in contraposition to actions like theirs.

Another remarkable Indigenous repertoire that might nuance and contextualize Del Mar's fencing opposition is the one created by Kumeyaay poet Tommy Pico. In Pico's poems, particularly those in their book *Nature Poem*, Pico aims to subvert the "Noble Savage" myth normally associated with Indigenous peoples in the account of their relations with nature.[39] Their book reckons with the journey of a young Native American poet who cannot write a poem about nature and, in the conflict, finds his own identity as an urban, queer, Native American writer. Water figures prominently in this journey, along with blood and drought, trees "slapped in the face," and an uncomfortable Pacific Ocean that appears in the protagonist's dreams with "massive deaths," "orange cliffs," and "fine sand." In the process of creation, circulation, and consumption of these poems, a gender nonconformant grammar of the beach otherwise emerges. This is a beach intimately silenced by colonial and capitalist practices of land dispossession and genocide but that has always, despite efforts to settle it, existed in flux.

The attention to beaches' unstable conditions for recreation, environmental problems, and related Indigenous water epistemologies have been recently complemented by academic, political, and mass media discussions on the Black Pacific. These discussions engage the past of erasures of Black people from the beach and are mostly centered in beachfront property land restoration.[40] The case of Bruce's Beach in Los Angeles, where oceanfront property was returned to the descendants of a dispossessed Black family by the California government, set an example for future social justice reparation projects along the Pacific, at least according to mass media accounts, until the family decided to sell their property back to the county of Los Angeles.[41] Discussions on the Black Pacific and cases like that of Bruce's Beach help to contextualize the racial legacies that inform beach disputes such as the one in Del Mar as well as the bodies, values, and practices deemed morally correct with respect to its uses for today and for tomorrow.

Indigenous repertoires, such as the ones here addressed, signal openings and alternative stories on the beach that, even while they have existed for so long, have been erased, neglected, or purposefully silenced. Indigenous repertoires are fugitive, embodied repertories that call for a shift in the mainstream, short-term, and extractive cultural understandings of nature as economic resource. They invite us to think-feel with the beach as an agentive being and not as a resource to put to work for economic, scientific, or leisured benefit. Kumeyaay repertoires here described are long-term formulations that imbue alternative temporalities in our relations with California beaches: not for now necessarily, but for tomorrow. They open up spaces for embodied memories that are not necessarily those that ensemble predictions work with but that are equally important in the movement toward viable, inclusive beach futures. These futures must include Indigenous and Black worldviews because their silencing and erasure keeps being a fundamental part of the problem.

Conclusion

Disputes over beach access bring to the fore the beach as a political and politicized terrain where multiple constituencies claim rights and engage in repertoires of future-making. In so doing, they serve to exemplify why beaches, like parks and plazas, malls or theme parks, matter as public spaces. Beaches are plural, fragile, and political ecosystems. Their political nature might be best apprehended in tourism contexts where shores and coasts have been engineered (politically and materially) to cater to the needs and desires of the privileged classes, often white and wealthy, often entitled and relating to the beach as a natural space already there and ready for their enjoyment. This does not mean, however, that the affects and cultural meanings, the uses and projections, of these spaces are the only ones existing.

Paying attention to beach access disputes and their specificities is an empirical avenue to show and understand the plural conceptualizations and uses of the beach that can happen simultaneously in time and space to those enunciated by the most privileged. Their representations and practices aim to fix and secure the beach for their individual interests, which often times coincide with those of mainstream, hegemonic culture in reading beaches as natural spaces to conserve and put to work for economic profit, mostly through tourism and recreation. Yet, there are also scientific readings of the beach, which also heavily rely on West-centric measurements and expert projections, serving as a dire reminder that beaches won't last long in their current configuration if not taken care of in more sustainable ways—and they will last even less long when hydrogeological processes seem to advance faster than predictions themselves, with sands and waters shifting even as humans engineer social and concrete walls to fix them. And finally, there are also those otherwise readings of the beach that enunciate the beach as a lived, changing, collaborative pedagogical space from which to learn about the political nature of cultural differences and other relational nonextractive understandings of sand and water.

The concept of *cultural repertoires* as "existing grammars that make social positions legible" has been a useful conceptual tool to make sense of the heterogeneous conceptions of the beach emerging from the Del Mar railway fencing proposal and beach access dispute. More broadly, it has served as an orientation into the toolkits at hand for individuals and collectives in their practice and imagination of beach futures in shared times of climate crisis and infrastructural expansion. Cultural fixation, scientific, and Indigenous cultural repertoires might be useful analytical tools to address beach politics beyond the beaches of Southern California. They might serve as tools to account for politics in other public spaces, too, such as plazas or parks, and as such, they are deserving of further conceptualization.

NOTES

1 Teresa Calderia, *City of Walls: Crime, Segregation and Citizenship in Sao Paulo* (Berkeley: University of California Press, 2000); Setha Low, *Behind the Gates: Life, Security and the Pursuit of Happiness in Fortress America* (New York and London: Routledge, 2003).

2 Matilde Córdoba Azcárate, *Stuck with Tourism. Space, Power and Labor in Contemporary Yucatan* (Oakland: University of California Press, 2020).

3 Ann Swidler, "Culture in Action: Symbols and Strategies" *American Sociological Review* 51, no. 2 (1986), 273–286, https://doi.org/10.2307/2095521; Michèle Lamont and Laurent Thévenot, eds., *Rethinking Comparative Cultural Sociology: Repertoires of Evaluation in France and the United States* (Cambridge: Cambridge University Press, 2000).

4 Lamont and Thévenot, *Rethinking Comparative Cultural Sociology*, 5–6.

5 Lamont and Thévenot, 4.

6 Mike Davis, Kelly Mayhew, and Jim Miller, *Under the Perfect Sun: The San Diego Tourists Never See* (New York: New Press, 2003).

7 Mimi Sheller and John Urry, *Mobilities: Places in Play, Places to Play* (London: Routledge, 2004), 1.

8 James V. Fenelon and Clifford E. Trafzer, "From Colonialism to Denial of California Genocide to Misrepresentations," *American Behavioral Scientist* 58, no. 1 (2014), 3–29; Lee M. Panich, "Archaeology, Indigenous Erasure, and the Creation of White Public Space at the California Missions," *Journal of Social Archaeology* 22, no. 2 (2022): 149–71. https://doi.org/10.1177/14696053211061675.

9 Waleed Hazbun, *Beaches, Ruins, Resorts: The Politics of Tourism in the Arab World* (Minneapolis: University of Minnesota Press, 2008); James Freeman, "Democracy and Danger on the Beach: Class Relations in the Public Space of Rio de Janeiro," *Space and Culture* 5, no. 1, (2002): 9, http://dx.doi.org/10.1177/1206331202005001002.

10 Ajantha Subramanian, *Shorelines: Space and Rights in South India* (Stanford, CA: University of Stanford Press, 2009); Bianca Williams, *The Pursuit of Happiness: Black Women, Diasporic dreams, and the Politics of Emotional Transnationalism* (Durham, NC: Duke University Press, 2016).

11 Christina Dunbar-Hester, *Oil Beach. How Toxic Infrastructure Threatens Life in the Ports of Los Angeles and Beyond* (Chicago: Chicago University Press, 2023).

12 North County Transit District (NCTD), *Proposed Fencing Project*, 2018, https://gonctd.com, accessed November 2023.

13 NCTD, *Risk Reduction and Feasibility Analysis: NCTD Trespasser Risk Reduction Analysis Report*, June 30, 2020, 6, https://gonctd.com.

14 California Coastal Commission, Protecting and Enhancing California Coast, https://www.coastal.ca.gov, accessed November 2023.

15 Luke Harold, "'Fence Fighters' in Del Mar Urge Their Neighbors to Oppose NCTD Fencing Plan," *Del Mar Times*, November 23, 2021, www.delmartimes.net.

16 Joe Tash, "NCTD Proposal to Fence Coastal Railroad Tracks Draws Strong Opposition at Del Mar Council Meeting," *Del Mar Times*, January 13, 2021, www.delmartimes.net.

17 Phil Diehl, "Transit District Downsizes Del Mar Railroad Fence, amid Continuing Protests," *San Diego Union Tribune*, October 17, 2021, www.sandiegouniontribune.com.

18 Phil Dielh, "Feds Still Sitting on Del Mar's Railroad Fence Decision," *Del Mar Times*, March 5, 2023, www.delmartimes.net.

19 Laura Place, "Del Mar Residents Sue Transit District over Bluffs Fencing Project," *Coast News*, April 15, 2022, www.thecoastnews.com.

20 Michael Taussig, "The Beach (A Fantasy)," *Critical Inquiry* 26, no. 2 (2000): 250, https://www.jstor.org/stable/1344123; Adam Keul, "The Fantasy of Access: Neoliberal Order of a Public Beach." *Political Geography* 48 (2015): 49–59, http://dx.doi.org/10.1016/j.polgeo.2015.05.005.

21 Heather Hope, "Del Mar Residents Fired Up against Railroad Fencing Project at Beach," *CBS8 News*, August 20, 2021, www.cbs.com.

22 Del Mar Union School District, "History of Del Mar," Del Mar Union School District, 2021, www.dmusd.org, accessed November 2023.

23 Daniel Quirk, "Stop NCTD from constructing train fence in 2020 on Del Mar Bluffs and in North County" *Change.org*, 2020, www.change.org, accessed November 2023.

24 Place, "Del Mar Residents Sue Transit District."

25 Hope, "Del Mar Residents Fired Up"; Chris Jennewein, "NCTD Moving Ahead with Fencing to Prevent Train Deaths on Del Mar Bluffs," *Times of San Diego*, January 28, 2022, www.timesofsandiego.com.

26 Dominic E. Reeve, Adrián Pedrozo-Acuña, Mark Spivack, "Beach Memory and Ensemble Prediction of Shoreline Evolution Near a Groyne," *Coastal Engineering* 86 (April 2014): 77–87, https://doi.org/10.1016/j.coastaleng.2013.11.010.

27 World Meteorological Organization, *Guidelines on Ensemble Prediction Systems and Forecasting* (Geneva, Switzerland: World Meteorological Organization, 2012).

28 Sarah E. Vaughn, *Engineering Vulnerability: In Pursuit of Climate Adaptation* (Durham, NC: Duke University Press, 2022).

29 Amitav Gosh, *The Great Derangement: Climate Change and the Unthinkable* (Chicago: University of Chicago Press, 2016), 37.

30 Stephanie Kane, *Just One Rain Away: The Ethnography of River-City Flood Control.* (Montreal: McGill-Queen's University Press, 2022).

31 Katherine McKitrick, *Dear Science and Other Stories* (Durham, NC: Duke University Press, 2021).

32 Maya J. Berry, Claudia Chávez Argüelles, Shanya Cordis, Sarah Ihmoud, and Elizabeth Velásquez Estrada, "Toward a Fugitive Anthropology: Gender, Race, and Violence," *Cultural Anthropology* 32, no. 4 (2017): 537–65, https://doi.org/10.14506/ca32.4.05.

33 Raymond Craib, *Cartographic Mexico. A History of State Fixations and Fugitive Landscapes* (Durham, NC: Duke University Press, 2004), 81.

34 Native Like Water, "Native Like Water. A program of One World BRIDGE Non-Profit 501c3," accessed March 7, 2023, www.nativelikewater.org/.

35 Native Land Digital, "Native-Land.ca," an online tool and app to help navigate Indigenous territories, languages, and treaties, 2023, https://native-land.ca/, accessed November 2023.

36 Todd Prodanovich, "Whose Coast Are You Surfing in San Diego? A Look at the Indigenous History of One of California's Most Iconic Surf Zones," *Surfer*, August 17, 2020, www.surfer.com.

37 Kym Trippsmith, ed., *Tribal Water Stories: A Compilation of California Tribal Stories, Position Papers & Briefing Papers in Conjunction with the 2009 Tribal Water Summit* (n.p.: California Tribal Water Summit Planning Team, 2010), https://water.ca.gov/.

38 Xilonen Luna Ruiz, *Kumiais: Homenaje a Gloria Castañeda Silva, cantante Kumiai. Pueblos Indígenas en Riesgo* (Mexico City: Comisión Nacional para el Desarrollo de los Pueblos Indígenas, 2008).

39 Tommy Pico, *Nature Poem* (Portland: Tin House, 2017).

40 Caroline Collins, "Black Americans Have Deep Ties to the Pacific but They Have Been Erased," *Washington Post*, February 8, 2023, www.washingtonpost.com.

41 Rosanna Xia, "Bruce's Beach Can Return to the Descendants of Black Family in a Landmark Move Signed by Newson," *Los Angeles Times*, September 30, 2021, www.latimes.com; Soumy Karlamangla, "The Debate on Bruce's Beach," *New York Times*, March 9, 2023, www.nytimes.com.

3

Experiences of Leisure, Governance, and Protest in the Coastline Spaces of Beirut

NADINE KHAYAT AND CLARE RISHBETH

Beirut, the capital of Lebanon, is situated on a peninsula on the eastern shores of the Mediterranean Sea. This spectacular location is intrinsic to Beirut's unique character, and many residents walk along the corniche, glimpse the sea from adjacent streets, and watch sunsets at Pigeon Rock. Even in times of severe and continuing social upheaval, these pleasures are free, and highly valued in the high-density urban grid of Beirut.[1] However, these experiences are also compromised through multiple acts of corruption, intentional neglect, and social exclusion. This chapter focuses on the social histories and cultural values of these conflicting dynamics.

Our purpose in sharing this complex story is to examine how coastal leisure spaces can be both "valued despite" and "undermined through" unstable governance in a postwar and high-migration context. Under-standing these tensions within the context of the Global South can offer insights into leisure spaces operating outside of rational strategies of design and care which are (comparatively) normalized expectations of local government in the Global North.[2] In this chapter, we use a case study structure, exploring three locations along the urban seafront, all primarily used for leisure purposes. These are chosen both to repre-sent different landscape typologies and for their use in highlighting key public space issues in Lebanon: privatization, gentrification, and pol-lution. We discuss how explicit or implicit state actions have shaped the common usage of these coastal locations. Our conclusions raise the possibilities of the resilience of leisure claims on public open space, al-ternative routes of intervention, and, above all, the enduring emotional draw of the sea in disrupted lives.

Introducing the Lebanese Context

Lebanon holds a historically strategic position on global east-west trade routes and in recent decades has reflected some of the geopolitical turmoil of the wider Middle Eastern region. Within living memory, this has included a protracted civil war (1975–90), ongoing conflict with Israel, and ongoing settlement of refugees from neighboring countries, most notably Palestine (1948 onwards) and Syria (2011 onward). Lebanon currently has the highest number of refugees per capita in the world, living in urban areas as well as in specific camp settings, with over 1.5 million Syrian refugees.[3] The country has many structural and civil divisions across ethnicity, religion, and income level and ongoing political insecurity within a postwar context.

For much of the twentieth century, Beirut was a cosmopolitan city and popular tourist destination, famous for its wealth, diverse cultural heritage, and spectacular location on a rugged peninsula (figure 3.1). The civil war years changed this perception and fundamentally damaged the physical fabric and spatial understandings of inclusion and belonging in the city. The city was split by the infamous "Green Line," a spatial and psychological demarcation line that separated the (broadly Christian)

Figure 3.1. Map of Beirut highlighting the case studies with inset of Lebanon. Source: Google Maps, adapted by N. Khayat.

east from (broadly Muslim) west neighborhoods. Thirty years on, this line is no longer a physical barrier, but it still resonates in social imaginings of city life, and lives on politically in the formal structuring of government along carefully designated religious identities. Lebanese citizens are acutely aware of the lack of political accountability, highlighting misuse of power and funds for personal gain often through deals on development, and the acute lack of investment in civic infrastructure.

Introducing Seafront Leisure Spaces in Beirut

This research project examines diverse experiences of seafront spaces in Beirut, with the specific aim of understanding how people and communities with different identities are able, or unable, to enjoy these unique locations. All residents of Beirut have experienced interruptions to normality within their lives. Middle-aged and older residents have lived in the city through a civil war and Israeli bombardment. Palestinian and Syrian refugees have been forced to leave their homelands. Ongoing and significant civil unrest and economic precarity impact the mundane patterns of daily life for everyone. What role does leisure and visiting seafront spaces play amid this ongoing demand for "getting on and getting by"? Who is supported in and who is marginalized from these spaces?

Often conceptualized as a privilege, leisure has numerous barriers for vulnerable populations living in poverty including time, transport, facilities, and discrimination, as well as fears of untamed landscapes, getting lost, and (primarily for women) sexual assault, and this can impact their quality of life.[4] However, even among those most disenfranchised, a small but increasing body of research looks at how spending recreational time in public open spaces and natural environments can support individual well-being and strengthen social networks.[5]

Looking at blue space and seafront typologies in the Global South offers an opportunity to counter most of the research regarding marginalization and leisure situated in the Global North, which mostly explores typologies of recreational spaces such as parks, forests, and public squares. Blue spaces positively affect people's health and well-being,[6] with seascapes in particular having a distinct effect on the human perception of movement and time, given its fluidity, movement, and intensity.[7] The sea is an important leisure destination in the Arab culture,[8] with seafront promenades termed a *corniche*. This specific typology of a recreational outdoor space combines movement with pausing, creating an extensive recreational zone on the water's edge, which can provide important social affordances and contact with nature.

This research contributes to understandings of socially and environmentally just cities and the need to align landscape planning with equity across race, ethnicity, and class.[9] These approaches include the role of governance and the extent and means by which people can influence policy, including participatory and activist actions. While debates on multiculture, ethnic diversity, and public space are relatively well established in the Global North,[10] it is important to rework both approaches and relevant identities specific to the Global South.

In the Lebanese context, Abir Saksouk-Sasso (2015), architect and planner, argues that it is the everyday practices, as opposed to the state governance of public space, that determine, produce, and sustain urban public space in the city. She notes that the state, the usual provider of public space, is either disinterested in or openly opposed to the provision of public space in Lebanon.[11] The role of protest is also distinctive. Authorities continue to act protectively against unrest after the Arab Spring (2010–11), a period when public space provided a catalyst and enabled civic protest across Tunisia, Egypt, Yemen, Iraq, and Lebanon.[12] Despite transformative changes in people's everyday lives from these actions, this did not lead to the disruption of the structure of governments.[13] Asef Bayat highlights that after the uprisings, people started to mobilize in public space for protests aimed at material improvements and new cultural and political norms, turning public space into a theater of contentions where conflicts of varying intensity involving residents and urban authorities are played out.[14]

Within the case study sections, we attend to the voices of users from diverse backgrounds as well as a range of stakeholder insights, enhancing understanding of the value of the seafront spaces, both past and present, and exploring some of the top-down and bottom-up impacts on the form and function of these spaces. In a following section we focus on significant collective actions that have offered some insight into how politically motivated vested interests have been and can be resisted on the Beirut seafront.

Methods

The mixed range of methods used in this study allow in-depth responses to specific sites but also an understanding of value across the whole seafront area and the relationship of this value to the city's geography and history. The aim is to accurately represent the extent of mixity of seafront spaces (by nationality, age, gender, and income) and to explore time-depth, both in terms of individual memories but also stakeholder accounts of place change and their drivers and impacts.

The three seafront sites are all adjacent to residential areas and were chosen to illustrate different typologies of open space and to exemplify a range of concerns related to social and environmental justice in the city (figure 3.2). The sites are

- Manara (also known as the corniche), a broad seafront walkway with a mostly rocky shore below, stretching about two kilometers tracing the edge of the peninsula;
- Zeituna Bay, previously a working harbor, now a high-end Marina edged with upmarket cafés; and
- Ramlet al Bayda public beach, the last remaining public sandy beach in Beirut.

Qualitative and quantitative methods were integrated focusing on understanding sites across different spatial and temporal scales.

Similar to the pioneering work by William Whyte (1980) and tested in urban design by Jan Gehl (2011), behavioral observations were recorded throughout the research, captured in the form of fieldnotes, photographs, and on-site sketching.[15]

The visitor profile of seafront users, and how this varied across the case study locations, was captured through an on-site questionnaire in 2019. Random sampling techniques were used and 441 users completed the survey. Data collected include demographic information, including where they had traveled from and details regarding their visit (social context, activities, length of stay, frequency of visits) as well as some short comments on their experience of the seafront.

This was followed up in 2021 by twenty-one semistructured interviews with seafront visitors. These were conducted on site with people spending

Figure 3.2. The case sites as they appear on the map (clockwise from left): Zeituna Bay, Ramlet al Bayda public beach, and Manara. Photos by N. Khayat.

time in two of the selected case study sites, Manara and Ramlet al Bayda, selected to represent a range of different user profiles. The interviews investigated user experience and gained an in-depth understanding of practices, preferences, and memories as well as the perception of other users of the seafront spaces and related tensions and pleasures.

To gain an understanding of top-down and bottom-up processes of change along the urban coast, seventeen semistructured interviews were conducted with stakeholders including government officials, activists, academics, designers, and nongovernment organizations (NGOs). These sought to determine priorities, timelines, and implications for policy, governance, activism, and the users of Beirut's seafront.

Finally, and gaining more significance due to the COVID-19 pandemic, digital archives and online visual media were recorded and analyzed. Particular attention was given to responses to historic material of the shoreline. Though this offers a somewhat biased sample of responses (people active on social media and interested in place), it allowed some more reflective responses to Beirut's leisure histories to emerge that would have been difficult to capture through other means.

Seafront Leisure Spaces and Environmental Injustices

Manara: Interrupted Leisure and the Encroachment of Private Interests into Public Spaces

The Manara (corniche), often seen as the iconic image of the Beirut seafront, runs adjacent to the highly dense residential area of West Beirut, which also clusters tourist hotels, the central business and banking area, and the American University of Beirut campus. While historic images of this length of seafront depict mostly uninterrupted views out to the sea, and access down to the rock foreshore, privatization from recent decades has resulted in a more mixed visual and physical access to the sea. This case study demonstrates some of the vested interests that can override the public good in the Lebanese context.

A landscape architect described the diversity, saying, "The corniche has multiple users depending on times and days: mornings for serious health-oriented, lunchtimes for recreational joggers and those enjoying a break while evenings are for gatherings of younger adolescents, couples and families with kids." She also described a change over time: "We used to have a relation with the sea, but this has been disconnected. People used to feel the coast, dip their feet in the water and that was a part of life; it was a right in the city and they had a history of doing this."[16]

Since the 1950s, the public's connection with the sea and practice of freely using the beach has incrementally been severed due to privatization and encroachment on the public domain enabled by the state. Lebanese laws including the Beirut Master plan, building laws, and the maritime zoning laws enshrine the right to seafront public property which cannot be sold, owned, or built on, according to the maritime zoning law, as and where the highest waves hit during the winter months including sandy, pebbled, and rocky beaches.[17] However, this law was superseded through decrees in 1966 and 1992 granting exceptions and building permits allowing public property to be sold and developed for tourism. The mechanisms for these exceptions have been the provision of building licenses, temporary public works licenses (that facilitate encroachment), and private titles for deeds close to the seafront.[18]

In the case of Manara, these developments are not necessarily large buildings but are beach zones partitioned for private use, members clubs, and private swimming facilities.[19] An investigative article in *Al Akhbar* newspaper names thirty illegally built developments on Beirut's seafront public property that do not have any permits. The result is that opportunities to swim off the corniche, to access the water's edge, or sunbathe by the shore are very limited especially for less well-off residents and those without Lebanese connections.

Zeituna Bay: Gatekeeping Diverse Uses of Spaces

The case study of St. Georges Bay, renamed Zeituna Bay in 2011, illustrates how commercialization and gentrification radically transformed a formerly open public space into a highly regulated semiprivate space primarily catering to the affluent. Here, the private sector invested large sums of money in the public realm, resulting in a high-end designed space and a pleasant marina experience. But the space has lost a lot of its publicness, and the survey showed a clear bias in user profiles toward people with higher incomes.

Before the redevelopment, St. Georges Bay was a public promenade around a small harbor, with the Port of Beirut on the eastern side of the bay. Beirut's commercial city center had been extensively damaged during the civil war, and in the period afterward, an extensive development process was instigated led by Solidere, a private investment company.

The role of Solidere in Beirut's reconstruction has been heavily criticized.[20] Though the original plans outlined "60 landscaped and open spaces comprising a large park, gardens, squares, pedestrian areas, quaysides and seafront," these were not delivered. The company devalued owners' property, bought them, and reconstructed a city center for the rich, paying very little attention to the provision of a high-quality urban realm.[21] The masterplan included St. Georges Bay, which was developed and rebranded as "Zeituna Bay, Beirut's finest leisure destination," a high-quality urban marina with expensive restaurants on the lower level facing the harbor.

Similarly, the Zeituna Bay development replaced everyday, inclusive, mixed public area with a more excluding ethos. Activities on the marina are strictly managed by a private company. Though there is public access to the decked marina walkway, many social activities are banned, such as dog walking, sitting on the grass, loud music, and fishing. Still, the fieldwork revealed a slightly more complex picture. Within Zeituna Bay, the decked walkway edging the water is the space with the most diverse set of users, especially during the weekends when lower income groups often have a day off work. During the weekends, it transforms into a space where Lebanese, Syrians, and migrant workers enjoy walking, taking pictures, drinking coffee, and feeding fish. A Syrian visitor valued this space, saying, "Its nicer at the weekend; it is relaxing, has a European feel to it; this is where I feel relaxed in Beirut, there is nature and urbanity around me, I like to sit here not along the corniche, I feel more welcome here."[22] However, most visits were fairly short as it is not easy to hang out without spending money. A Lebanese visitor noted, "It should be more open to the public; it is also very expensive and should be more accessible to people."[23]

The users and uses of this highly gentrified area have further changed in recent years as a result of the actions of political uprisings of 2019; we explore this recent dynamic later in the chapter.

Ramlet al Bayda Public Beach: Pollution and Less Valued Seafront Visitors

Ramlet al Bayda, or "white sand" in Arabic, is increasingly notorious as one of the most polluted sites in the city.[24] It is the last remaining sandy public beach in Beirut and is popular for bathing, but users are exposed to a range of health hazards due to the presence of open sewer outflows; a report by the National Council for Scientific Research named Ramlet al Bayda a high-risk area unsuitable for swimming due to harmful bacteria and fecal remains.[25]

Despite its dilapidated and highly polluted state, a high number of people use the beach especially during the summer months. There is a diverse mix of users, notably from lower income backgrounds, with a higher proportion of Syrian visitors. Groups of men can be seen running, playing football, volleyball, and backgammon, while groups of families and friends enjoy swimming, picnics, playing music, and just hanging out. Many beach users are aware of the high level of pollution on the site, and some restrict their activities to walking rather than swimming. A user commented on the "trash polluted, unhealthy environment." She said, "I live in front of Ramlet al Baida and the color of the sea changes, sometimes it is dark blue-greenish-black. I assume it's all the oils and hazardous things that are dumped in the sea over time."[26]

Ramlet al Bayda is a clear example of environmental inequalities being actively perpetuated. In Beirut, a few public sites are well-managed while others are left to decay.[27] The Ministry of Public Works, the Municipality of Beirut, and the governor are responsible in varying degrees for the maintenance and upkeep of public spaces in Beirut, but they have not closed off open sewers that have been polluting the site for almost a decade, and overlaps of responsibility have meant that there is not a clear framework of accountability.

The Ministry of Public Works has outsourced Ramlet al Bayda's beach maintenance to an NGO, Operation Big Blue Association. The NGO advances social justice in the city through the ethics of caring and repair, small acts of organizing cleanups and adding fresh coats of paint, which can contribute to a sense of well-being.[28] They claim that the lack of funds restricts more extensive management actions. Big Blue Association makes some of their income from beach users through a small kiosk selling refreshments and hiring out umbrellas, chairs, and loungers on the public space. This approach is not valued by all users, with frequent complaints that the rentals are overpriced considering the socioeconomic demographic of users.

There has been an organized call from practitioners, academics, and activists for Ramlet al Bayda to be better protected as a shared open public space.[29] The focus, however, has been the more urgent issue of resisting overdevelopment rather than the incremental and less dramatic story of pollution levels. In 2015, the municipality issued a building permit for a significant tourist resort called Eden Bay despite the current zoning and building regulations.[30] With public backlash reaching higher levels including protests on the beach, the Municipality of Beirut assigned Ramlet al Bayda as a zone under study, halting further development for two years. In a postwar Global South context like Beirut, there are often multiple and interlinked threats to both the existence and the quality of public open space.

Collective Actions for Seafront Spaces in Times of Political and Social Turmoil

These case studies offer insight into environmental and social equity in Beirut, a city that exemplifies many tensions that shape postwar, high-migration cities in the Global South. The postwar years in Lebanon have been shaped by ongoing, increasing public anger about the lack of accountability of the state and the influence of corruption on decision-making. Much of this focuses on the eroding ethos of the public good and the *publicness* of the city. In terms of spatial planning, there is extremely minimal resident involvement in planning processes and an absence of formal participatory processes. However, our fieldwork documented methods and tactics where power was seized or negotiated to protect seafront leisure spaces and to resist state actions. This section sets out two different timescales of bottom-up interventions, both responsive to the specific political turmoil of Lebanon but also with resonances to mechanisms of change in other Global South contexts. The first is an immediate act of protest prompted by a national crisis; the second demonstrates a more strategic cross-sector collaboration. Both were underpinned by a collective belief in a right to access seafront spaces.

The first focus is on the nationally and globally significant uprising of October 2019 (termed the Thawra) which took place during the research fieldwork and had a particular impact on public spaces in Beirut, including many of the seafront spaces.

On October 17, 2019, hundreds of thousands of citizens took to the streets in cities around Lebanon calling for a change in the political system and for representation under the weight of a growing economic crisis deeply rooted in corrupt public practices.[31] Crowds of activists and ordinary city residents, across usual sectarian divisions, gathered in public spaces across Beirut, seizing these as venues for connection, revolution, and debate. Newsreels were shared around the world captured noisy (largely peaceful) occupations of city spaces which had been neglected or underused long-term, including abandoned cultural buildings, city plazas, and the seafront.[32]

Protests, peaceful sit-ins, and human chains of peace formed by linking arms were organized on Manara to demand unequivocal and equitable access to public spaces as a "right to the city." This strategi-cally photogenic public venue was chosen by protesters as a location that envelopes the whole city and is valued by diverse city inhabitants who united across class, nationalities, ethnicity, and religion.[33] Many of the slogans and spokespeople focused on resisting a neoliberal creep in city planning which favors private investments in large-scale tour-ism projects instead of safeguarding the accessibility of seafront pub-lic spaces. Protesters shrouded parking meters along the seafront and sprayed the phrase "We Are Not Paying." Bou Aoun, architect and public space activist, recalls, "I got involved on the seafront because it is en-tirely hijacked by real estate development, tourism, business tycoons it is completely insane, most of our border is on the seafront but if I live in Beirut I cannot go to the beach."[34]

The protests also focused on the gentrification and intrusive manage-ment style of Zeituna Bay, and the "grey area" of whether this was legally defined as public space. Groups occupied the wooden decking level of the marina to assert their right to occupy this space as public property and resist the illegal acquisition of seafront public property. Two cre-ative actions were hosting an open-for-all breakfast on the decking and screening, for free, the movie *V for Vendetta*, a 2005 thriller in which a vigilante attempts to bring down a fascist government.

An activist with a primary role in the protests on Zeituna Bay explained:

> Zeituna Bay happened when we were blocking roads. We did the break-fast, we removed the placards prohibiting activities—music, picnics—and that was it. We decided to stay and camp on Zeituna Bay. A lawyer had given us advice. "If LSF [Lebanese Security Forces] don't kick you out it is a win because you reclaimed the space. If they do kick you out and the LSF arrives on site you will be fined. They will have to write a reason why on the fine, for example 'camping on public property.' Then we can file a lawsuit because they [LSF] admitted it is public land." In this instance, the LSF did not fine the protesters but the management of Zeituna Bay did release a circular admitting to the "publicness" of the wooden public walkway, which essentially meant everyone is free to practice the space without prohibitions. Since then people were able to fish, sit on the grass, play music and bring their dogs.[35]

In response to the occupation of Zeituna Bay, the management company released a statement acknowledging the "publicness" of the wooden walkway, resulting in a small win for public space activists in Lebanon.

The second example is the Civil Campaign to Protect the Dalieh of Roauche.[36] Lebanon has a strong NGO and higher education sector, and in certain times and places, strategic coalitions have formed to intervene on public space and planning issues.[37] A history of collaboration between activists and academics relating to the Dalieh headland provides a hopeful example of how it is possible to seize the initiative in terms of reclaiming and restoring landscape quality and access rights.

Dalieh is a limestone headland at the far west of the Beirut, at the end of Manara and just north of Ramlet al Bayda beach. The most naturalistic stretch of the Beirut coast, the headland functions as a visitor attraction for viewing the coast and taking trips out in leisure boats, and it has a small fishing community. Dalieh has been subject to various transgressions including the threat of an illegal land sale, undermining the heritage of the site.[38] Academics, activists, and practitioners launched a civil and public campaign to protect the access to and the ecological and heritage value of this site in March 2013.

Abir Saksouk-Sasso, a member of the campaign, activist, and co-founder of Public Works Studio, explains her motivation and the process involved:

> I started out as an individual. I was working on the Dalieh campaign, and I was working as an active participant because I had already researched Dalieh. This provides you with a passion for taking this work though; also I am someone who used to go to Dalieh and Ramlet as a kid. The campaign developed and did what it had to do, and I started working through Public Works. While many organizations work on the seafront, we are expanding this research to encompass all public land in Lebanon and how the state has historically dealt with this through privatization to make money to settle its deficits, especially during this economic crisis.[39]

While still operating outside a functioning democratic planning system, Dalieh provides an example of how decisions about seafront spaces can be more collectively made. The space is now more highly protected because this campaign raised the profile of Dalieh among academics, practitioners, and activists who formed a coalition. No attempts at building, threats to development, or harassment to fishermen have occurred since. Key factors for this achievement appear to be a purposeful solidarity and collective effort between activist organizations with the relevant skills and capacity. An extensive network of connections was also important. The campaigners were able to spread information about the site in major media outlets, including holding an international design competition through the order of Engineers and Architects in Lebanon, raising the profile of the value of this landscape.

Conclusion

In this chapter we have briefly explored some of the sociopolitical dynamics of seafront spaces in Beirut, a postwar Global South context. Through these case studies, we have traced some of the pressures on the *publicness* of these spaces: tourist development, gentrification, pollution, and loss of heritage. These threats can equally impact other forms of public spaces, such as parks, but within all these cases, the presence of the sea and the special qualities of sea-edge recreation heighten the past and potential losses of recreational opportunity to the residents.

The first point to note, connecting both the personal and the political, is the irreplaceable value of a sea edge. Encountering the sea, walking along the corniche with sand under your feet and views to the horizon over Manara, is a well-being resource for mind, body, and soul that is fundamentally different from any other kind of leisure activity. On one hand, in a highly segregated city, it is a leveler. Whether refugee or citizen, school child or professional, these pleasures connect at a human level and are (travel costs aside) free to enjoy. On the other hand, this value is also economic. In a postwar society with extensive redevelopment challenges and a beautiful natural and cultural environment, tourist and "high-end" commercial development makes sense politically. But given poor regulation, this development is also open to motivations of greed and opportunities for corruption, maximizing the accrual of assets to those with power. The intrinsic value of a sea view is also a spatially finite resource and, therefore, the most highly threatened of public goods.

Environmental justice in a city like Beirut is fundamentally linked to access to and diverse meaningful use of leisure spaces. The case studies here show that the informal daily recreation of city residents is not highly valued by high-status stakeholders such as the municipality. It is easily dismissed when there are opportunities for development profiteering, or where the lack of a political will and complexity of legal responsibilities mean pollution or neglect is ignored. Our research focused on the experience of marginalized communities who had fewer leisure options.[40] While a few did benefit from some of the public space improvements (e.g., some visitors to Zeituna Bay enjoyed the better facilities), a much more common story was that tourist-focused developments deterred use by low-income residents, and the refugee-background users of the seafront were far more likely to be present in the lower profile, low facility locations such as Ramlet al Bayda or to put up with the pollution on the one remaining public beach.

Another important lens for understanding seafront recreation in the Middle East is the enduring cultural value of seafront spaces through precarious and disruptive times. In this chapter, we focused on the impact of the October 2019 political uprisings, but our fieldwork also revealed the vital importance of the different locations throughout the COVID-19 pandemic and in the following financial collapse in Lebanon. The impact of the lockdown meant that for a while, the corniche reverted to a local neighborhood recreational resource, a place of connection and solidarity. During the economic crisis (still ongoing), and with other leisure options extremely limited and electricity supply intermittent, the corniche has become very busy. While this has positive aspects to it, the intensity of the crowds highlights the poor maintenance of these spaces. Investment for maintenance and improved design is key to the future of the seafront to accommodate the well-being needs of Beirut residents.

Finally, we looked at the context of place change relating to seafront spaces in a country where there is no formal public participation in local governance. The case studies demonstrate the positive role of stakeholder coalitions for public space advocacy and the intermittent success of this advocacy in Lebanon. Academics, NGOs, and activists connect with different communities and differently claim rights to public space and the protection of heritage and raise the need to address pollution. While there are still challenges for the long-term resilience of these voices and the representation of these non-middle-class, globally connected circles, each successful step models a more egalitarian future and provides a prototype for future action. Here we can see long-term strategic advocacy alongside reactive and proactive public protest working together as a catalyst for diverse public participation in the use and future direction of Beirut's seafront.

NOTES

1 Beirut has one square meter of green open space per person. This compares to forty square meters of public space per person recommended by the World Health Organization.

2 The terms *Global North* and *Global South* are a method of grouping countries based on socioeconomic variables. *Global North* broadly comprises developed countries such as North America and Europe while the *Global South* has evolved to broadly refer to countries and regions in Latin America, Asia, Africa, and Oceania, in particular those which are low to middle income. The term is used not only to highlight an economic profile but also the issues of political instability or culturally marginalization that impact many of these locations.

3 "UNHCR Lebanon at a Glance," UNCHR: The UN Refugee Agency, Lebanon, accessed August 15, 2020. www.unhcr.org.

4 Iwasaki Yoshitaka, "Leisure and Quality of Life in an International and Multi-cultural Context: What Are Major Pathways Linking Leisure to Quality of Life?" *Social Indicators Research* 82 no. 2 (2007): 233–64.

5 Clare Rishbeth, Farnaz Ganji, and Goran Vodicka, "Ethnographic Understandings of Ethnically Diverse Neighbourhoods to Inform Urban Design Practice," *Local Environment* 23 no. 1 (2018): 36–53; Estella Carpi, Jessica Anne Field, Sophie Isobel Dicker, and Andrea Rigon, "From Livelihoods to Leisure and Back: Refugee 'Self-Reliance' as Collective Practices in Lebanon, India and Greece," *Third World Quarterly* 42 no. 2 (2021): 421–40.

6 Sebastian Völker and Thomas Kistemann, "Reprint of: 'I'm Always Entirely Happy When I'm Here!' Urban Blue Enhancing Human Health and Well-Being in Cologne and Düsseldorf, Germany," *Social Science & Medicine* 91 (August 2013): 141–52.

7 Mike Brown and Barbara Humberstone, *Seascapes: Shaped by the Sea* (Florence: Taylor & Francis, 2015); Anna Ryan, *Where Land Meets Sea: Coastal Explorations of Landscape, Representation and Spatial Experience* (London: Taylor & Francis, 2012).

8 Waleed Hazbun, *Beaches, Ruins, Resorts: The Politics of Tourism in the Arab World.* (Minneapolis: University of Minnesota Press; 2008); Christine Delpal, "Une Promenade de Bord de Mer: La Corniche de Beyrouth," in *Reconstruction et Réconciliation Au Liban: Négociation, Lieux Publics, Renouement Du Lien Social,* eds., Chawqi Douayhi and Eric Huybrechts (Beyrouth: Presses de l'Ifpo, 1999), 187–207; Abdullah Addas and Clare Rishbeth, "The Transnational Gulf City:

Saudi and Migrant Values of Public Open Spaces in Jeddah," *Landscape Research* 43 no. 7 (2018): 939–51.

9 Julian Agyeman, and Jennifer Sien Erickson, "Culture, Recognition, and the Negotiation of Difference: Some Thoughts on Cultural Competency in Planning Education," *Journal of Planning Education and Research*, 32, no. 3 (2012): 358–66.

10 Leonnie Sandercock, *Cosmopolis II: Mongrel Cities in the 21st Century* (London: Continuum, 2003); Patsy Healey, "The Universal and the Contingent: Some Reflections on the Transnational Flow of Planning Ideas and Practices," *Planning Theory* 11, no. 2 (2012): 188–207.

11 Abir Saksouk-Sasso, "Making Spaces for Communal Sovereignty: The Story of Beirut's Dalieh," *Arab Studies Journal* 23 (2015): 296–318.

12 Nadine Sika, "Contentious Activism and Political Trust in Non-Democratic Regimes: Evidence from the MENA," *Democratization* 27 no. 8 (2020): 1515–32.

13 Asef Bayat, *Revolutionary Life: The Everyday of the Arab Spring* (Cambridge, MA: Harvard University Press, 2021).

14 Asef Bayat, *Life as Politics: How Ordinary People Change the Middle East* (Stanford: Stanford University Press, 2013), 8.

15 William H. Whyte, *The Social Life of Small Urban Spaces* (Washington, DC: Conservation Foundation, 1980); J. Jan Gehl, *Life between the Buildings: Using Public Space*, (Washington, DC: Island Press, 2011).

16 Stakeholder Interview, January 2019.

17 The maritime public domain, in the Lebanese legislation (the decree 144/a, dated June 10, 1925), is defined as being the aquatic port and the coast to the farthest distance the waves could reach in winter, as well as sandy and pebble beaches. The seawater ponds and marshes linked to the sea are also a part of the maritime public domain.

18 Jadaliyya has written about the campaign and compiled their articles; see "The Civil Campaign to Preserve the Dalieh of Beirut," *Jadaliyya*, www.jadaliyya.com.

19 Zbib Mohamad, "Who is Stealing the Sea," *Al-Akhbar Newspaper*, December 5, 2012.

20 Stakeholder Interview, January 2019.

21 Saree Makdisi, "Laying Claim to Beirut: Urban Narrative and Spatial Identity in the Age of Solidere," *Critical Inquiry* 23 no. 3 (1997): 661–705.

22 Questionnaire respondent, July 2019.

23 Questionnaire respondent, July 2019.

24 Suzanne Baaklini, "Water Quality Report: Two thirds of Lebanese beaches are not safe for swimming," *L'orient le Jour*, July 22, 2022.

25 Baaklini.

26 On site interviewee, August 2021.

27 Julian Agyeman, *Introducing Just Sustainabilities: Policy, Planning, and Practice* (Zed Books, 2013).

28 Setha Low and Kurt Iveson, "Propositions for More Just Urban Public Spaces," *City* 20 no. 1 (2016): 10–31.

29 Scott Preston, "The Untouchable Hotel," *Executive*, April 16, 2018, www.executive-magazine.com.

30 Preston.

31 Mona Fawaz, Mona Harb, Howayda Al-Harithy, and Ahmad Gharbieh, "The Beirut Blast: A week on," *Beirut Urban Lab*, October 8, 2022.

32 Cynthia Bou Aoun, "Reclaiming Public Space and Its Role in Producing the Revolution," *Legal Agenda* (blog), January 8, 2020.

33 Khayat, Nadine, and Clare Rishbeth. "Exploring Urban Co-presence and Migrant Integration on Beirut's Seafront," *Migration Studies* (2023).

34 Stakeholder interview, July 2021.

35 Stakeholder interview, September 2021.

36 "The Civil Campaign to Protect the Dalieh of Beirut."

37 See the work of the Beirut Urban Lab, https://beiruturbanlab.com; and Public Works Studio, https://publicworksstudio.com.

38 "The Civil Campaign to Protect the Dalieh of Beirut."

39 Stakeholder interview, October 2021.

40 "The Civil Campaign to Protect the Dalieh of Beirut."

4

Beach Time

The Politics and Moral Order of Argentina's Urban Beaches

MARIANO D. PERELMAN

In Argentina, the warm climate draws thousands of visitors to beaches, particularly in the province of Buenos Aires, from mid-December to the end of February. I will refer to this moment as *beach time*, with its sports, late nights, temporary friendships, meetings, and goodbyes. The practices of beach time are not entirely ruptured during the rest of the year; however, they do differ. The beach, as an open public space, has its rules of interaction and its forms of appropriation. Beach time has a profound impact on urban order. It shapes it through the formation of legal and illegal labor and consumption practices as well as through the creation of distinct spaces for political and social conflict that become inherent to the moral organization of the city. Beach time enhances individual and societal behaviors, expanding and amplifying their influence on the urban order. By examining beach time, we gain insight into the political nature of the urban moral order and the centrality of beaches in it.

As is common in many parts of the world, Argentine legislation safeguards unrestricted entry to the coast. It asserts that all waters, shores, beaches, and entrances to them, in their entirety, are "public goods" and the "inalienable" heritage of all civilians. Access and use of the beach must be ensured to all individuals who desire to avail themselves of such areas. Nevertheless, the access and utilization of beaches are subject of dispute, and certain groups lack the capacity to access and use the beach as much as others do.

This chapter examines two cases where forms of living, working, and expressing values come into conflict with beach time. The first case centers on the beach as a place of work and commerce. This study focuses on ambulant vending in Mar del Plata, Argentina's primary summer tourist city.

It demonstrates that while beach vending is not illegal, vendors still must negotiate their presence with other commercial and political actors during beach time. In doing so, they deploy micro-manipulations of norms, site-specific agreements with other beach actors, and shifting moral justifications around work on the beach as a public and commercial space.

While the first case focuses on work and commerce on the beach as a way of living that structures disputes over perceptions and practices of what is the legitimate use of public space, the second case sheds light on the development of contested moral values around cultural prejudices and women's bodies during beach time. I trace the popular example of the *tetazo* protests in 2017. The *tetazo* protests began with a complaint in a coastal city in the province of Buenos Aires when two local women requested that the police arrest two other local women going topless. Both groups of women were residents of the city and had prior rivalries based on sexual and political orientation. The complaint was followed by a police intervention to prohibit toplessness which was facilitated within the repressive policies of the national government of Mauricio Macri (2015–19). Subsequently, feminist mobilizations against the police intervention, known as the *tetazo protests*, were organized in various cities throughout Argentina. This case exemplifies how a local and private dispute over gender that played out *at the beach* and *during beach time* became elevated to a national discussion on gender morality with repercussions well beyond the physical space of the beach.

Doing the Season: The Beach as a Place of Work and Commerce

In the province of Buenos Aires, beach time is a particular moment when the beach is transformed. During the summer months, the beach becomes more socially and economically valuable. It becomes a place of work and commerce wherein different collectives negotiate the larger urban moral order.

Beach time generates in Argentina a particular *cultural setting* where time and space come together to make the beach an *events-place*.[1] The physical features and warm weather of the beach are integral, as are the social and political procedures that transform beaches into vacation destinations providing access to commerce, relaxation, and political and labor activism.

The establishment of beach politics, the creation of place-making and embodied space, and the political transformation of space are not solely determined by government action or legislation but also by the utilization of laws and the interconnectedness of actors and their agencies. Thousands of people go to the beaches for vacation during the summer—beach time—so it is common to find in Argentina's beaches street vendors seasonally selling various products such as food (ice cream, churros, candy, corn), CDs, kites, and clothing (tights, T-shirts, pants, hats, socks, among others). While many beach vendors reside in the cities along the Atlantic coast, most vendors *go* to the beach to work only during beach time. The season for these vendors implies a particular physical movement from the cities (Buenos Aires) to summer places such as Mar del Plata and their presence must be negotiated with local actors, including local beach vendors, for whom the beach is a yearlong space for both work and recreation. Negotiation processes involve (in)formal space governance and often lead to disputes over what constitutes a morally acceptable and legitimate way of living, being, and working at the beach as a public space.

While ambulant vending on the beaches is not prohibited by the state, each municipality can regulate or even ban it.[2] Further, the seasonal flow of tourists and vendors changes the way the beach is regulated. Oftentimes, regulations are dormant during winter and they are enforced during summer, when the beach becomes a socioeconomical value-producing place. To gain access to sales, vendors must negotiate their presence and activities with security personnel, employees of beach concessionaires, and groups that manage street commerce often through violence.

In summer, beaches are spaces of commerce and socioeconomic opportunity for vendors, both yearlong residents and seasonal workers. However, for beaches to become spaces of opportunity, it is necessary to generate practices of affinity with other vendors, negotiate spaces of possibilities (which imply arrangements), and construct legitimate ways of using the beach as a public space. Whether they reside in Mar del Plata or travel there for the season, vendors must negotiate, albeit through varying methods.

For those working seasonally in Mar del Plata, *doing the season* contrasts with vending on the streets of Buenos Aires. Juan sells throughout the year on a regular bus route in Buenos Aires. To work, Juan, like other vendors, establishes relationships and builds a specific territory. Meanwhile, Raúl peddles various items on trains, working during off-peak hours. Unlike Juan, he negotiates his route with other vendors and maintains a fixed schedule and route. The materiality and constrained spatiality of urban transportation infrastructure limits the practices of these vendors.

In the city, Juan, Raúl, and numerous vendors with whom I conducted fieldwork actively establish interpersonal relations to sell their products which help somehow transcend these spatial and material constraints. These connections persist throughout the selling process and outside of their work hours, with gatherings occurring at bars, in neighborhoods, and in churches. In these gatherings, information regarding products for sale is exchanged, friendships with security personnel are formed, selling prices are negotiated, and city territories are divided among vendors.[3]

The vendors with whom I conducted fieldwork claimed their *freedom* to work in the city and did not pay the police. "We organize ourselves, talk to each other, but do not pay the police."[4] Freedom and avoiding police payment (*arreglos*) are core moral principles within the daily urban practices of Buenos Aires city vendors and are part of their way of living. Spending time with other vendors is also significant to their way of life. Nevertheless, due to the beach's configuration and various stakeholders' existence, vendors must modify these learned practices and renegotiate their way of living if they want to make a living selling on the beach too. Like working on the city streets, working on the beach necessitates frequent negotiations with other actors. The trip to coastal cities, however, requires entering a domain governed by a logic distinct from that governing the city, necessitating the creation of *arreglos*. Traveling from the city to the beach entails, for example, acquiescing to alternative spatial constructions.

Juan spent ten seasonal years in Mar del Plata before 2013 when he decided not to go there anymore. He was tired of dealing with the police, other vendors, and the *mafia* that heavily influenced all aspects of work in the city and frequently demanded payment for labor. "Over there, everything is controlled by the mafia," he told me. When I asked him what he meant, he replied, "They ask you for money to work. It is the *barra brava* of the Varela club; you must pay depending on whether you want exclusivity or not for sales."[5]

In Buenos Aires, the term *mafia* commonly refers to street vending. As I learned through my fieldwork, middle-class discourse frequently associates the mafia with lower-class forms of organization.[6] Juan took the mafia accusation differently—for describing the *barras*, who were labeled as mafia by the media,[7] and other groups who demanded money from him to work on the beach. Juan viewed the vendors as victims and not as mafiosos. In his opinion, there was no alternative but to pay since these individuals had control of everything and held connections with the mayor and the police. This idea was shared by numerous other vendors who operate in Buenos Aires and undertake the *season* in Mar del Plata. Paying for doing the season was considered part of the order that governed the beaches. While it was possible to avoid payment for work in the city, on the beach, it remained a legitimate and much needed practice. Paying to work was an integral part of ambulant beach vending as a circuit of commerce.[8]

How products are offered is also different on the beach than in the street.

> You must make yourself visible among many people. You must use your body, you must walk among the people, you have to be listened to, and you have to let them know what you are selling. You have more time, anyway, not like on trains. You have all day; you must walk and look at people. Get attention and make yourself visible. It is easier on trains. People on the beach are doing other things. Also, many people get annoyed when you yell in their ear [*laughs*]. You must watch out for the sun and walk on the sand, but it is worth it.

Doing the season implies rethinking the spatial negotiations of working as well as a shift in the practice and allocation of moral values. The various agreements between actors facilitate the practice of vending. However, they also aid in the activity's criminalization and informal regulation. Engaging in vending necessitates reconsidering the spatial arrangements of labor and ethical principles. The illegalities connected to commerce form part of a political governance system and of the circuit of commerce.[9]

Beach time has effects on those who live in Mar del Plata. First, vending becomes a full-time occupation for them; second, they must negotiate their space on the beach with larger groups of people that travel there to make the season; third, they need to justify their presence, unlike throughout the year. Esteban is a summer beach ambulant vendor in Mar del Plata and works as a plumber for the rest of the year. "When summer comes, it is a problem. People come from outside and do not respect the beach. We are here all year round. We take care of the beach, that drugs are not sold, that they [visitors] are not stolen [from]. . . . We provide both a sale and a social service," he says.[10] Esteban seeks to argue publicly a legitimate justification for his presence and occupation of the beach as a working space. He knows he could be expelled. Therefore, Esteban tries to present himself as a legitimate worker.[11]

Thus, in addition to negotiating with law enforcement and local government, he prioritizes correctness and safety for beachgoers. "The community here has known me for years and trusts me," he explains. For Esteban and numerous other vendors residing in Mar del Plata, there is a coconstructed relationship between time spent on the beach and time spent working in the same town during the rest of the year. "During long weekends, I sell items on this beach to earn some extra money. However, this particular beach differs from others," says the vendor.[12]

For instance, a group of vendors in Mar del Plata requested to be recognized as a public service. They reported that they organized themselves to ensure that the beaches were free of "drugs," "lack of control," and "poor quality products" sold by other vendors. Rafael confirmed, "We are providing a service to the people."[13] Presenting themselves to the authorities as public servants, local vendors construct their presence and legitimacy against other vendors.

While seasonal vendors negotiate their presence with the police and other actors, it is also important to construct themselves as essential actors in the urban landscape. According to Javier, vendors "bring joy to the beach with our unique way of offering products. Imagine a beach without vendors? Impossible! We are an integral part of the beach, providing food, clothing, cigars, and music."[14]

The commercial circuit of the beach season differs from that during the off-season as well as of that established in the city. As demonstrated by Juan's case, spatial limitations (streets of Buenos Aires versus beaches of Mar del Plata) impact his willingness to negotiate. Street vendors are associated with their work in Buenos Aires. Additionally, beach time configuration, even in a large city like Mar del Plata, is constructed based on the off-season, as seen in Esteban's case.

Vendors capitalize on beach time as an opportunity to earn money and have fun. Ramiro mentioned that beach vending, unlike that in Buenos Aires, offers additional benefits. "It's enjoyable to work, eat, and appreciate the beaches," he said. "But you work," I pointed out. "The beach is a world apart. It can be exhausting with the sand, heat, and sun. However, there are the friends, the sea, and the parties," he retorted.[15]

Vending as a form of living comprises aspects of daily life such as selling, spending money, and moral values. Working on the beach entails participating in power dynamics that produce illegal practices, such as negotiating with the concessionaire, police, bars, and other vendors; selling "informal" products; not paying taxes; and so on. Additionally, it influences how actors behave on the beach and shapes their approach to selling.

The influence of beach time extends beyond its duration. Notably, regulations governing activities on the beach instill vendors with relational ways of existing in that space. It is a precarious occupation of the public area, and the earnings garnered can be saved for critical periods of the year. Beach time demonstrates that their livelihood in Buenos Aires is not akin to the mafia style of the *barras*. Consequently, it is a multifaceted experience that enables us to grasp the activity beyond the physical space of the beach. In Buenos Aires, moral arguments for the legitimate use of public space are often based on the experiences of street vendors and the organization of beach activities.

Tetazos: The Beach as a Gendered Place of Political Amplification

As public spaces, beaches are locations where conflicts arise and are resolved. In Argentina, during the summer season, the beach emerges as a highly politicized place. As Jennifer Bidet and Elsa Devienne show, the politicization of beaches is not a new phenomenon.[16] The beach is at the center of contemporary political and economic struggles: the rising sea levels resulting from climate change and urbanization; the increasing privatization of the coastline; the transformation of beaches into leisure spaces that create tensions with other industries and has resulted in the appropriation of this area by some social groups at the expense of others; and, as in the previous case, public uses for producing socioeconomic value.

In Argentina, beach time presents a critical period for political activity, including formulating security policies, antidrug operations, and political campaigns. However, I wish to introduce an alternative approach to politics by highlighting the story of a group of women who utilized their time at the beach as a political opportunity to challenge conventional lifestyles and their associated moral values through the exposure of a private conflict. Beach time is a pivotal moment in constructing urban order through public behavior and interactions. Such interactions and the management of politics generate "proper" conduct based on moral norms rather than legality.

In the middle of the season, on January 28, 2017, a conflict emerged in a coastal city in the province of Buenos Aires, when two women were reported for sunbathing topless. Two other women had contacted the police, who intervened and requested the topless duo to either cover themselves or depart the area (figure 4.1). The police presence prompted a lively debate among the women (and those on the beach), all of whom lived in the mid-sized city and knew one another for having divergent political affiliations and social views. "I'm astounded that she reported the girls," remarked a friend of one of the accusers during a conversation with me one afternoon. "It was understood there were tensions due to their sexual orientation and political affiliations with the Kirchnerist party, a center-left political group. Nonetheless, I'm surprised they involved the authorities." After briefly pausing, she added, "In a small city like this, everyone knows each other."[17] The complainants turned past private grievances into a more extensive moral, social, political, and public debate on gender morality in which naked women on the beach were deemed immoral. Both groups of women were challenging the hegemonic male-dominated urban moral code, albeit making different claims of how women could be in the beach as a public space.

Both groups of women seized the opportunity presented by summer crowds and media to voice concerns and register complaints. As stated by Mariana Garzón Rogé, actors' practices are embedded within a social world that requires interpretation and cannot be viewed as automatic or thoughtless.[18] Through contextual interpretation, the public denunciation gained strength. The ones who publicly denounced the women exposed

Figure 4.1. Police and women who were the subject of reports discussing on the beach. Source: Diario La Nación.

their neighbors to the local community. In contrast, the women who were denounced expanded the issue from a simple complaint to the persecution of women and their rights. A few days later, politicians discussed the case while feminist organizations conducted numerous *tetazo* protests across several squares in Argentina denouncing the sexualization of women.

In the conflict at hand, when the police requested that women cover up or leave the area, they cited Article 70 of the Criminal Code of the province of Buenos Aires (Chapter III, Against Public Morality and Good Customs). This section states, "Individuals will face punishment if they offend public decency through obscene acts, words, drawings, or inscriptions. The penalty will be doubled if the offense was committed in a location where public events or performances take place or against individuals who are worshiping, elderly, mentally ill, women, or children."[19]

The conversation between several women (and some men) who were reported to the police officers (of both genders) highlighted the significance of moral arguments around the body in constructing public order: "You're not going to tell a man to cover up if he's half-naked and wearing a bikini top," one of the women stated. "This is sexism," she added. "We are acting following the laws" and "this is a law violation," responded the police.[20]

By using the norm (written or unwritten) and a supposed moral order, the women tried to criticize others for deviations from the correct way of living. Actors employ public discourse to position themselves and act beyond the scope of the "urban order." They deliberate on an amalgamation of moral practices that shape lifestyles, including sexuality, political beliefs, neighborhood disputes, and even leisure activities such as going to the beach. Urban order is a result of public forms of living that are imposed and reinforced by people. These norms are well-known and consistently managed, with the expectation to adhere to the law and establish a relationship with it. Just a few days after this conflict, a criminal judge in the city made a declaration that "going topless was not a crime." But the importance of the denunciation was contextual.

This conflict rose from a local level to a national one. News reports presented the *tetazo* protests as a conflict centered on the appropriate way to behave on the beach and the legitimate use of public space. As described by Martín Boy, the former Minister of Security for the province of Buenos Aires, Cristian Ritondo, stated that the legality of going topless was irrelevant.[21] The concern was whether the action negatively impacted family morality in public spaces. As a result, time spent at the beach was seen as a means of reinforcing proper behaviors and maintaining the urban moral order. The beach becomes a contested space over the use of bodies and the morally just use of public space rather than a mere local conflict. It is crucial to examine this dispute from a gender-focused and politically situated perspective.

The women who reported the issue may be categorized as *moral entrepreneurs* or *crusaders*, according to Howard Becker's theory.[22] However, I aim to expand upon Becker's viewpoint by emphasizing the significance of normative struggles. People criticize and make critical use of norms before attempting to impose them. For instance, the complaint against women going topless deployed a moral argument to resolve a political conflict.[23] Moreover, the women managed to use this complaint to highlight gender inequality. Hence, it is apparent that the norm's substance is less significant than the actors' ability to create so-cial and moral distinctions among groups. The complainants and the accused utilized the beach as an amplifying space-time to generate and control discrepancies.

The conflict on the beach gained national attention quickly due to the convergence of beach time and women's mobilization. This phenom-enon was made possible by an inward and outward dual movement. Lo-cally, managing neighborhood conflicts allowed for the publicity of the beach conflict; nationally, it prompted a public discourse on women's rights. Both groups of women used their time at the beach to voice their demands and forge connections: one group denounced the repressive policies of Macri's government against minorities while the other pub-licly denounced the feminist movement.

The women who spoke out sought to change the perception of the conflict; the women's collective performed political work to make the act equivalent to gender-based violence and gender inequality. The women modified the scale as a form of vindication and conducted a nationwide *tetazo* protest called the "*tetazo* against police action and discrimina-tion." The inscription of social processes in specific grammar is crucial to perpetuating conflicts. If denouncers addressed norms and moral-ity, then the denounced women ensured the dispute was incorporated into feminist demands such as "On the sovereignty of women's bodies," "Against patriarchy," and "Against police violence."

The complainants denounce their neighbors on the public scale (the beach) based on their reading of the context. The complainants accuse their neighbors on a national scale based on their reading of the context. Laws such as the Sexual Education Law, the Equal Marriage Law, and the Gender Identity Law, which were enacted during the Kirchner govern-ments (2003–15), as well as certain specific inclusive policies connected to human rights, were questioned in different ways by the national gov-ernment of Macri (2015–19) and the government of the province of Bue-nos Aires of María Eugenia Vidal (2015–19).

In both the denunciation on the beach and the reaction of feminist groups, the social and urban order is at issue. Beach time enabled this articulation and led to unexpected changes initiated by the actors them-selves. These changes ranged from disciplining women who exposed their breasts to the justice system affirming that it was not a criminal act.

In addition, it gave the conflict substantial outreach and supplied human rights groups with tools to pursue retribution. The beach's politicization facilitated the denunciation at a municipal scale and beyond. The utilization of denunciation, police intervention, and social reaction showcase the mobilization of moral arguments pertaining to law and rights. The societal deployment of law and its breaches generate solidarities, maxims, and customs that will eventually obtain a "living right." This requires us to reject a simplistic, binary understanding of politics that divide into legal or illegal and formal or informal categories.[24] Beach time in Argentina proves that the beach as an embodied space that is both subject to and creator of moral orders is a good starting point in this rejection.

Beach Time and Its Political and Moral Grounds

My ethnography demonstrates that beach time is a paradigmatic cultural setting for comprehending (in)formal politics through microinteractions, negotiations, and utilization of law. The production and maintenance of unequal forms of living—based on access to resources, uses of spaces, and ways of imposing moral values—are produced by actors in everyday practices. The cases presented in this chapter reveal the beach, and particularly beach time, as public and politicized realms that engender societal and moral differentiations propelled by and through political interventions. The behavior exhibited on the beach has high visibility and serves political, social, and moral purposes. The beach accentuates and amplifies forms of living and conflicts. It does so particularly during beach time when the physical space of the beach becomes a favorite ground for moral and political debate for citizens and authorities alike.

In both cases explored in this chapter, the beach enhanced social and political processes. In the first case, addressing the prohibition of vendors and the requirement for labor negotiation, the construction of "illegal" ways of being in space requires negotiations and produces unequal ways of using space experienced differently by seasonal and resident vendors in Mar de Plata. Vendors equally profit from the situation by having the opportunity to earn money, yet each needs to deploy different practices and strategies, which in some cases involve reverting their moral orders of what is appropriate to do to be able to work. In the second case, beach time permits actors at the beach to denounce publicly and, by changing the conflict's scale, reinforce larger urban moral order on gender (or contest it). The act of denunciation and *tetazo* protests are active forms of body politicization of, on, and by the beach. Beach time promotes political involvement and facilitates the widespread reach of local disputes.

The use of laws as a means of legitimized arguments during beach time has a significant impact on the institution of formal and informal legislation throughout the year. The example of vendors utilizing legal and moral arguments to distinguish themselves from other informal actors, such as barras and other vendors, is a testament to this. The case of the *tetazo* protests exemplifies how women initiate large-scale social processes to denounce and counteract regressive politics on their bodies. Processes occurring at the beach have significant impacts on larger urban governance.

The politicization of beaches in Argentina and beyond is at the center of contemporary social and economic struggles. Beaches are crucial to shaping forms of living and working, including social interactions, behaviors, and recognition of others. This occurs in the daily interactions of flesh-and-blood people who struggle to impose ways of life based on questioning and accepting behaviors. Today, work, recognition, and the imposition of an urban moral order have become central themes for beachgoers.

NOTES

1 Setha Low, *Spatializing Culture: The Ethnography of Space and Place* (New York: Routledge, 2016); Antonádia Borges, *Tempo de Brasília: Etnografando Lugares-Eventos Da Política*, Coleção Antropologia Da Política 21 (Rio de Janeiro: Relume Dumará, Núcleo de Antropologia da Política, 2003); E. Valentine Daniel, *Charred Lullabies: Chapters in an Anthropography of Violence*, Princeton Studies in Culture/Power/History (Princeton, NJ: Princeton University Press, 1996).

2 According to national law, provinces possess jurisdiction within three miles from the territorial sea line (twelve miles). However, provinces delegate functions concerning coastal beaches to municipalities, which are responsible for management and control over these areas. For employment at the beach, municipalities may demand credentials such as a declaration of products to be sold or proof of residence.

3 Mariano D. Perelman, "Vender Nos Ônibus. Os Buscas Na Cidade de Buenos Aires, Argentina," *Tempo Social* 29, no. 1 (April 15, 2017): 69, https://doi.
org/10.11606/0103-2070.ts.2017.124659.

4 Raúl (vendor), comment to the author, Buenos Aires, September 2013.

5 Juan, in conversation with the author, Buenos Aires, April 2013. Football fans who
engage in violent practices are commonly referred to as *barra, hinchada*, or *barra
brava*. These groups base their actions on concepts of *aguante* (strength) and
honor. *See* Pablo Alabarces, ed., *Hinchadas*, Colección La Mirada Antropológica
(Buenos Aires: Prometeo Libros, 2005); Pablo Alabarces, José Garriga Zucal, and
María Verónica Moreira, "El 'aguante' y las hinchadas argentinas: una relación
violenta," *Horizontes Antropológicos* 14 (December 2008): 113–36, https://doi.
org/10.1590/S0104-71832008000200005. Club name has been changed.

6 According to Pacífico and Fernández Álvarez, "'Mafia' is a frequently used term in
the Argentinean mass media when discussing street vending. The term suggests
that street vendors are controlled by illegal organizations run by entrepreneurs
who take over public space and engage in human trafficking." Florencia Pacífico
and María Inés Fernández Álvarez, "Nunca Mafia: Experiencias de Vida y Formas
de Organización de Vendedores Ambulantes En Espacios Públicos," in *Bajo Sos-
pecha: Debates Urgentes Sobre Las Clases Trabajadoras En La Argentina*, ed. María
Inés Fernández Alvarez, Sandra Ileana Wolanski, Dolores Señorans, Florencia
Daniela Pacífico, Carmina Pederiva, María Paz Laurens, María Silvana Sciortino
et al. (Buenos Aires: Callao Cooperativa Cultural, 2019).

7 For example, see David Cox, "El lado oscuro de las barras bravas en Argentina:
Una mirada a la violencia en el fútbol," *CNN Español*, June 14, 2018, https://cnne-
spanol.cnn.com.

8 For Zelizer, a circuit has the following elements: a distinctive set of social relations
among specific individuals; shared economic activities carried on by means of
those social relations; common accounting systems for evaluation of economic
exchanges, shared meanings that people attach to their economic activities; a
well-defined boundary separating members of the circuit from nonmembers with
some control over transactions crossing the boundary. Viviana Zelizer, "Circuits
of Commerce," in *Self, Social Structure, and Beliefs: Explorations in Sociology*, ed.
Jeffrey C. Alexander, Gary T. Marx, and Christine L. Williams (Berkeley: University of California Press, 2004), 122–44.

9 Mariano Daniel Perelman, "Mercados informales y violencia(s) en Buenos Aires,"
Antropolítica Revista Contemporânea de Antropologia, no. 50 (December 22,
2020): 34–61.

10 Esteban (vendor), conversation with the author, Mar de Plata, July 2021.

11 The necessity of a legitimate way of using public space has been a constant on
my work with ambulant vendors and waste pickers in Buenos Aires. See Mariano
D. Perelman, "Disputas En Torno al Uso Del Espacio Público En Buenos Aires,"
Caderno CRH 31, no. 82 (2018): 87–98.

12 Esteban (vendor), conversation with the author.

13 Rafael (vendor), interview with the author, Mar del Plata, January 2021.

14 Javier, interview with the author, Buenos Aires, September 2014.

15 Ramiro, interview with the author, Buenos Aires, July 2015.

16 Jennifer Bidet and Elsa Devienne, "Beaches of Contention," *Actes de la Recherche en Sciences Sociales* 218, no. 3 (June 15, 2017): 4–9.

17 Mariela, conversation with the author, Buenos Aires, March 2017.

18 Mariana Garzón Rogé, "Aprendices en un país extranjero: Notas para una historia pragmática," in *Historia pragmática: Una perspectiva sobre la acción, el contexto y las fuentes*, ed. Mariana Garzón Rogé (Buenos Aires: Prometeo, 2017), 24.

19 "Del régimen contravencional," accessed September 22, 2020, https://normas.gba. gob.ar/documentos/ZBOPDhkV.html.

20 "Topless de tres mujeres desata un escándalo en Necochea," YouTube video, 00:02:50, January 28, 2017, https://www.youtube.com/watch?v=RJW3Dgs6eXc.

21 Martín Boy, "El Cuerpo Limitado En El Espacio Público: Conflictos En Torno al Género y La Sexualidad," in *Sociología y Vida Urbana*, ed. Verónica Paiva (Buenos Aires: Teseo Press—FADU/UBA, 2021), 245–68.

22 Howard Becker, *Outsiders: Hacia una sociología de la desviación* (Buenos Aires: Siglo Veintiuno Editores, 2010).

23 Mariano D. Perelman, "Dollars, Pesos and Planes: Reconstruction of Class Borders in the Second Government of Cristina Fernández de Kirchner (2011–2015)," *Dialectical Anthropology* 45 (2021): 253–73.

24 Emilia Schijman, *A qui appartient le droit?: Ethnographier une économie de pauvreté*, Droit et Société: Recherches et Travaux 33 (Paris: Maison des Sciences de l'Homme, LGDJ, 2019).

5

The Post-Political Beach

Conceptual and Empirical Explorations in Greece and Austria

SABINE KNIERBEIN AND CHARIS CHRISTODOULOU

Enclosure of public beaches, maritime coasts, and lakeshores has been an issue in the Mediterranean and central Europe for the past decades, as recently exemplified in Greece and Austria. Urban and regional beaches have come under pressure especially through trends of privatization, tourism, and real estate development as well as through other exclusionary and luxury encroachments. However, the right to public waters and access to public and partly state-owned lakes and seashores are part of both European Union (EU) policies and the Austrian and Greek legal systems, albeit in diverging ways. This chapter investigates the privatization of coastlines around Thessaloniki and privatized access to lakeshores and natural bathing areas with a special focus on the case of the Attersee in Upper Austria. We relate these cases on different analytical levels to wider trends in what we coin the *post-political beach*, that is, the phenomenon of post-politicization expressed by increasingly difficult public access to beaches and a wider depoliticization of basic rights to enjoy nature.

Beach, in this sense, is not just a depiction of a sandy and tropical strip of nature next to blue waves, but it is

— a social construct and a category of social experience;
— a geography of everyday life, difference, and marginalization; and
— a sphere of the commons, a public good, and a collective imaginary.

We are interested in the interrelation of these social, cultural, and political theory dimensions. Beaches relate to free access to and unobstructed use of places of well-being, social encounter, sports, education, leisure, and the experience of being directly exposed to nature, although the places are often urbanized and sometimes represent harsh environments.

Beaches have been coined as a particular public space where *the social* often plays out in very distinctive ways. When analyzed in socio-economic terms, beaches have been cultivated to cater to different class backgrounds, yet research has also witnessed increasing constraints on the open use of beaches, mainly pushed through privatization for the sake of tourism. This has affected how the public generally views the beach as a place of convivial urban experience and as a common good, a fact that points to the shifting political character of beaches.

We explore the hypothesis that beaches have become a crucial arena of post-politicization around which new struggles to repoliticize the beach may take place. We situate the debate of the privatization of beaches in those accounts theorizing the post-political condition that seek to study how democratic (public) institutions and representatives of democratic systems act; how institutional disregard (uncare) might lead to neglect in the public protection of the commons; and finally, how activists, civic initiatives, and NGOs may or may not try to reinstate democratization by stating *the political*. Our key questions are: How democratic is the beach? How political is it? And is it even post-political?

In what follows, we first introduce the theoretical context for the analysis of (post-)political features of beaches (the theoretical beach), followed by two case studies in Greece and Austria (the empirical beach), both geopolitical contexts pertaining to the EU, and its supranational legal frameworks. The case studies complement each other, yet they carry one main difference: in Greece, free access to maritime beaches is guaranteed by the national constitution, while in Austria, such a basic constitutional right to enjoy and access nature does not yet exist. The analysis of the beach in Greece starts from an empirical grounding of depoliticization, privatization, and failing governance of public beaches; the analysis of the beach in Austria is initiated by detecting a post-political condition on a structural level, then examine the policy level, respective acts of resistance, and potential paths of repoliticization. While we intend to develop an analytical spectrum that may help scholars analyzing beaches to carry out future research into the complex (post-)political features of beaches, some insights from a loose comparative debate of both cases will be shared. Our research methods were mainly qualitative content analysis of daily and weekly press sources, official public announcements, and institutional framework and of policy documents and our own fieldwork observations and procedural analysis of policy decisions and planning processes.

Post-Political Beaches?

Thinkers analyzing processes of post-politicization tend to engage with a more recent political philosophy approach: postfoundational theory. This theory "engages with the matters of the grounding of society, its 'fundamentals,' interrogating if and how universal principles of equality and freedom are institutionalized following moments of liberation" and emancipation.[1] A key aspect of postfoundational thought is the concern that "the political re-enactment of equality can only emerge because of the inevitable contradictions of a social order which presupposes equality but simultaneously disavows it."[2] This situation has also been termed as post-political condition.[3] Postfoundational theory helps to explain why and how contemporary representative democracies like those present in Austria and Greece allow for or even promote beach privatization trends with irreversible marks on maritime and alpine landscapes (depoliticization) or how they eventually politicize the quest for civilians' right to access and use public waters (repoliticization). As follows, we recap translational efforts of the scientific debate on the post-political condition to urban studies, planning theory, and public space research and on space as a mode of political thinking.[4] Thereby, we conceptually contextualize our research endeavor and carve out an approach to study the post-political beach.

For two decades, a return of the political has been identified in urban studies providing empirical evidence or conceptual advancements,[5] for example, with a focus on territorial governance,[6] specters of radical politics,[7] public space,[8] or lived space.[9] These refer to empirical findings or political theories that state that in public space, the political may be enacted through the everyday spatial practices of publics using this space.[10] Scholars have criticized how the shift toward governance implied a displacement of issues "from arenas of public debate and decision making into closed networks of elite representatives and technical experts" and consequently produced "glaring democratic deficits" as issues of pressing public concern "become sheltered in shadowy forums comprised of select groups of influential stakeholders."[11]

During the past few decades, the advent of a so-called depoliticization of public space has been witnessed, a process in which, for example, the public realm is strategically and symbolically shaped for certain target groups with specific consumption power. Certain phenomena of social exclusion are often shown as closely connected to processes of redesigning the material arrangements of public plazas, streets and parks, or beaches. Such place-making practices favor a selective conception of well-funded target groups, as they tend to exclude low-income and other marginalized groups and individuals, over an inclusive approach to constantly changing urban publics that is open to everyone.[12] These debates point to the crucial role of space in processes of post-politicization: space, as Mustafa Dikeç has it, is a mode of political thinking.[13] The use of space in these political theories "is not haphazard; . . . different spatial imaginaries inform different understandings of politics."[14] In this sense, beaches may also inform a mode of thinking about how politics or the political can be explained.

Inspired by the concept of radical social imaginary as put forward by Cornelius Castoriadis,[15] we approach the beach as a collective imaginary, deeply political, in which people have the freedom to express themselves and act autonomously. For Castoriadis, the social imaginary is not a mere reflection or veil but the framework through which human beings mediate and enact reality.[16] Every society lives within a symbolic system and develops an image of the world. In every instance, this image is created by producing meaning, which is important for collective life: "The imaginary has no flesh of its own. It borrows its substance for the rational, the investment of fantasy, the ascription of value, the symbolic, everything subordinated to effectiveness."[17] Post-politicization then occurs with the subordination of the collective imaginary to capitalist profit, which derives from alienation from the institutional procedures.[18] As Carolina Crijns views it, Castoriadis's idea of institution relates to the individual, who is necessarily socialized and embodied in collective society—therefore, social autonomy implies and presupposes individual autonomy.[19] By exercising their autonomy, people actively participate in the remaking of society. This means that society is self-instituting by its ability to self-reflect and distance itself from its own imaginary to reinterpret and recreate it. Because society recognizes itself as the source and origin of its own existence, society can undo what it has created.[20] Our focus on the collective imaginary of the beach hence implies that the beach constitutes a distinct post-political spatiality, a (new) terrain for the neoliberal project's acts of depoliticization, and hence a spatial imaginary, a field of potential repoliticization, too.[21]

In this regard, theorizations of a positive relation between public space and democracy need to be put within the wider context of the evidenced ambivalent depoliticization of the politics of public space. Considering beaches as geographies of everyday life, or even as public space, shows that these relational conceptions of space are both material and social. Coining the beach, then, as post-political frames a sociopolitical critique that points to the fact that "increasingly representative structures of democratic governance" have witnessed a deep affront on the public sphere and the public realm. The result is an increase in depoliticization of public space politics.[22]

Depoliticization of Struggles to Access Beaches in Austria and Greece

With this conceptual repertoire, we inspect phenomena of depoliticization concerning reduced public access to beaches—maritime in Greece, alpine in Austria. The Greek case focuses on the interplay of post-political governance and the misuse of the collective imaginary in the metropolitan area of Thessaloniki along Thermaikos Bay. The Austrian case refers to a general sociopolitical debate on the defacilitated public access to lakes across Austria and provides evidence from studying the Attersee in Upper Austria, a federal state of Austria.

Depoliticized Planning along Thermaikos Bay

In Greece, the relation of the people to the coasts belongs to a collective imaginary, contested and emancipatory in history. Hence, the right of people to freely access, use, and enjoy the seaside is protected according to the National Constitution.[23] Simple use concessions offered by the state to private enterprises are legally allowed exclusively for activities that serve the people, provide public recreation, or are for specified functional purposes.

The uses of public beaches in Greece, however, have recently turned out quite contrary to their constitutional conception due to neoliberal reforms and managerial changes that were enacted to combat the so-called Greek debt crisis (2009–18), along a continuous interplay of democratic and postdemocratic politics.[24] These happened within the legislative framework of the Economic Adjustment Programs imposed on the country by supranational entities, namely, the European Commission (EC), the European Central Bank and the International Monetary Fund, that rearranged governance at new scales within and beyond the state, consolidating a postdemocratic regime.[25] These programs basically induced a series of loopholes for exceptions and exemptions from the national public planning and the till-then valid legislation for beaches and seasides. When, for instance, the beach bears an emblematic quality, state property is allowed to be managed entrepreneurially, as an exception. Also, so-called Special Spatial Plans have been introduced as a new tool for focused developments that are not necessarily in harmony with Local Spatial Plans, the latter usually advanced through official planning and public consultation.

The objective of these reforms was to transform these commons into private property of the state, which could then take advantage to capitalize beaches' value supposedly for the benefit of economic growth. A range of Greek beaches have been dispossessed by means of national state acts to cater to private interests. Beautiful or scenic public coastal sites famous for their public use value (common good) have been commodified by exploiting their exchange value (public asset).

The waterfront of Thessaloniki is the city's most striking feature and most important public space, a significant common for all residents. The entire open sea unites people as a place of radical freedom nested in the collective imaginary (figure 5.1). It is this common imaginary that the recently forwarded Seafront Special Spatial Plan (SSSP) invokes as it attempts to superficially institute in the public discourse a new version of the future of the city's seafront along forty kilometers on Thermaikos Bay. In hegemonic fashion, it redefines Thessaloniki's seafront and beaches to a vast extent irrelevant to everyday experience. The redevelopment includes urban public plazas, solid waterfront promenades, and heritage sites as well as preserved *natural* landscapes such as rocky seashores, sandy beaches, saline flood plains, and transitional areas of

Figure 5.1. Thessaloniki urban seafront: Changing sea surface and view to Olympus mountain. Photo by Charis Christodoulou, 2021.

urban sprawl. It comprises sections pertaining to seven adjacent municipalities along the coastline of the metropolitan area, home to one million inhabitants. The inauguration of this new phase of urban planning, deregulation, and scaled governance (via the SSSP) came about as a means to accomplish a continuous bike route along the bay to integrate new aspects of mobility and well-being that proved to be a Trojan horse of privatization and commodification to take over most of the coast. Looking more closely at the planning and governance processes, one finds exceptions were made with planning procedures and new actors were involved at the local, national, and supranational levels. Thessaloniki joined the one hundred Resilient Cities Network (pioneered by the Rockefeller Foundation) and, in 2017, gradually set forth *Resilient Thessaloniki: A Strategy for 2030*, a document which adopted and reinforced the relevant jargon paving the way for private encroachments in and through public discourse. Alongside these strategies, the need was declared to simplify the governance of multiple official actors via a new governance system, "a platform for collaboration of stakeholders."[26] The resiliency strategy for the waterfront was paired with a framework redevelopment plan in 2018 that revisited the waterfront vision and formulated key steps "to catalyze the redevelopment of various . . . underutilized sites along the waterfront as a means to achieve social, economic and fiscal benefits."[27] The latter was elaborated by a consulting firm with the involvement of the World Bank.[28]

In November 2017, the local consultation process started with public presentations of the plan by privileged speakers.[29] Consecutive impromptu collaborations between the project manager, local authorities, and the local chamber of engineers (a quasi-state authority) were additionally set up. The final version of the SSSP was published by the Region of Central Macedonia on November 18, 2021, following an overtly restricted consultation period with poor results in which local citizens were not invited to participate. The plan passed through local authorities' elected councils by consensus to support urban development as most municipalities huddled together to retrieve planning gains based on the commodification of the beach instead of reclaiming its public access use as a common good. Exceptionally, one municipality out of seven, Kalamaria, opened the consultation process locally, which gave way to a wide dissent against the privatization of the local coast that would deprive the inhabitants from directly accessing the water for swimming, fishing, sailing, and scouting. Despite this critique, local and national media simply echoed the hegemonic discourse of the necessity for the waterfront plan with the objective of creating future urban sites of interest.

Free access to beaches in Greece is provided by the constitution, and everyone believes that this constitutional right will remain. However, constitutional rights protecting public access to Greek beaches continue to be ignored and private encroachments are not juridically banned; instead, they are facilitated. Via vitrine politics and shadow agreements among privileged stakeholders who consented to the SSSP plan along Thermaikos Bay, proceedings resulted in increasing social injustice, no recognition of people's relationship with the sea, and, finally, actual disengagement of local inhabitants from their lived democratic space.

Depoliticized Beach Governance around and beyond the Attersee

Austria is famous not just for its natural alpine landscapes and skiing resorts, but also its abundance of lakes (around twenty-five thousand). Examples of privatization of beaches, land leaseholds with houses built close to public waters, and natural bathing spots are omnipresent and wide ranging. This is because "the Austrian Law of Water Rights Act regulates the use of public waters—and allows for a non-restricted freedom of use—yet it does not regulate the access to these waters."[30] Evidence shows that 80 percent of the beaches at the Wörthersee in Carinthia are in private hands. The situation is similar for the Attersee in Upper Austria, where 76 percent of beaches are private.[31] The latter takes up an area of around forty-six square kilometers, while its shoreline comprises about fifty-three kilometers.

The investigation of the privatization of Attersee's beaches highlights the paramount role of the Oesterreichische Bundesforste (OBf), a previously public national entity responsible for Austrian forests. OBf was privatized in 1996, shortly after Austria joined the European Union in 1995. As part of an attempt to reduce national debts, *OBf* was then reinstituted as a private stock corporation (OBf AG) pertaining 100 percent to the Austrian national state.[32] The government at the time declared it a strategy to enhance the effective, revenue-oriented real estate transactions on the land that was now managed by the OBf AG.[33] During the successive conservative far-right national government (2000–2005), the idea was developed to shift eleven lakes from public to private property of the state. This move was considered to circumvent constitutional regulation, which protects the selling off of Austrian forests.[34] Effectively, in that period, eleven lakes, including the Attersee, comprising approximately ten thousand hectares of lakes and adjacent property plots were transferred to the property of the OBf AG with the goal to reduce national debts.[35] With this shift to state action based on private market economy principles, the OBf AG was required to pay a sum of approximately three billion Austrian Schilling (approximately €218 million) for the transaction to the state, which meant it was pressured to more intensively exploit the real estate value of the Austrian national forests, in particular the eleven lakes and the surrounding land plots.[36] Since its existence as a stock corporation, the OBf AG was capable of drawing enhanced profit out of no less than 10 percent of the country's territory.[37] The organization also considers itself the largest Austrian lake manager, as it takes care of seventy of the bigger lakes; at twelve of these lakes, they supervise forty-five natural bathing spots, ten of which are at Attersee.[38]

There has been high demand from the private sector to gain direct access to the public lakes, so the OBf AG sensed new business opportunities: they turned public beaches into commodities by leasing out beach land to private agents.[39] Via its privatized entities, the national state leases attractive public properties with direct access to the water, restricting its overall accessible lakeshore even more. At the Attersee, a dispute over another pertinent spot of land owned by OBf AG unfolded in 2019 around a public beach resort named after the OBf (Bundesforstebad), which had been let previously to the local municipality in Weyregg. Between 2019 and 2022, OBf AG privatized another fourteen-meter strip of lake access within the premises of this public beach by turning part of it into privately rented property (see figure 5.2). OBf AG argued that by making profit out of this strip, they would be able to enlarge the lido's sunbathing area and build a new food kiosk for public provision.[40] In 2019, however, the state parliament of Upper Austria declared a new state goal, free access to all lakes in Upper Austria in the federal district's constitution,[41] which proved ineffective. Around the lido, a public conflict between an activist group, the municipal authority, and the OBf AG emerged, which led to a protest petition being submitted to the Austrian parliament. This again did not stop or change the course of commodification of public waters and processes of post-politicization of beaches at Attersee. Another crucial concern relates to the increase in secondary residences affecting tourist and natural areas around Austrian lakes in general, and the Attersee in particular.[42] While statistics cover first residences, the growing number of secondary residences around the lake remains unclear and mainly beneath the radar of public planning regulation.[43] Because of its natural beauty, the Attersee region has seen the construction of secondary residences since the nineteenth century, with a steady increase since the 1960s. Two spatial planning regulations have been used in the past to guide decisions for or against secondary homes. Whereas the 1972 Spatial Planning Law for Upper Austria made it difficult to allow for a new zoning specifically for secondary residences in green areas, the 1994 Spatial Planning Law for Upper Austria facilitated such types of interventions as it explicitly prevented building secondary residences in areas zoned for housing. This produced the side effect that secondary residences were established in other zones, often those of emblematic character.[44] At Attersee, secondary residences are located at the lakeshore plots or on nearby hillside sites, thus creating a negative impact on the lake's accessibility and contributing to a trend of splinter development settlement patterns. Once direct lake access and emblematic views are at play, the real estate market value of

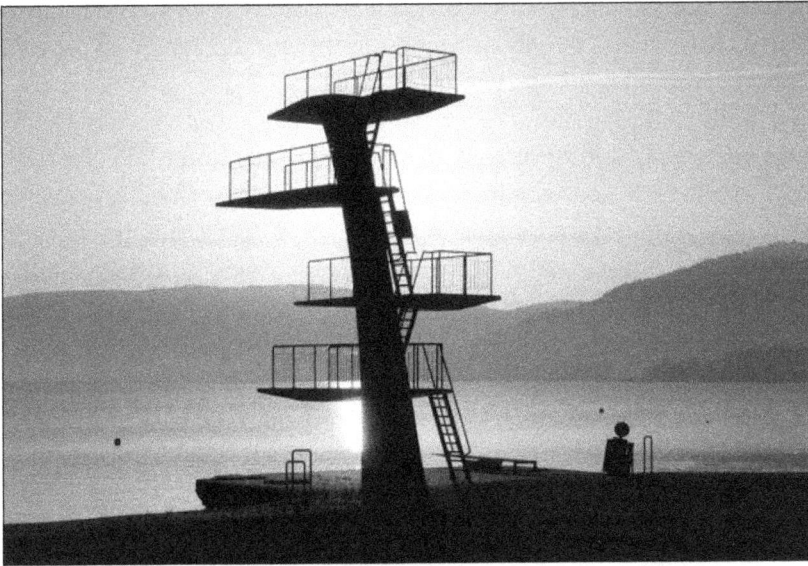

Figure 5.2. Diving station in Seewalchen at the Attersee in Upper Austria. Source: Creative Commons by Verleimnix Verleimnix, GFDL and CC-BY-SA-2.5, https://commons.wikimedia.org/w/index.php?curid=1565315

the site increases exponentially. These political-economic dimensions of planning policy ultimately trigger the process of real estate–driven privatization of beaches at Attersee.

Beyond Depoliticization

It is our aim here to show that both Austria and Greece, specifically Upper Austria and Thessaloniki Metropolitan Region, and local authorities of both Attersee and Thessaloniki have entered in a post-political phase when it comes to caring for beaches and access to public waters. We have emphasized the neoliberal policies leading to the privatization of beaches. In both countries, the absence of an outraged civil society contributes to an overall post-political condition around beaches. It is our objective to study how democratic (public) institutions and representatives of democratic systems act, how institutional uncare might lead to decline and neglect in the public protection of the commons, and, finally, who might be an agent to trigger the repoliticization of beaches as a mode of political thinking, where *the political* can be reinstated. This section develops around this last question and asks for civil society's form of response, resistance, and recovery.

In Greece, post-political governance of active citizens makes a complex interplay of consensus and orchestrated contestation.[45] People took part in a hype of coastal participation while they remained ignorant of the environmental transformation, eminent privatization, public land grabbing, and (essentially) shrinkage of public space, or they silently consented as they had been deprived of democratic expression in staged debates and were thereby kept out of public and civic decision making. With only one exception, local communities and citizens were not involved. In other places, a couple of citizens engaged in direct actions to claim part of the waterfront and instantly took ownership, yet these events often ended up with exclusionary or elitist practices such as festive events and superficial cultural performances.[46] The interested public was never directly informed that this new phase of urban planning is decisive for the future of the coastline as a whole. In fact, the SSSP spots beaches where development is uncertain and redefines them by providing new development rights largely uncontested. Until 2023, people widely remained ignorant of the eminent "shrinkage of public space" along the coast, while its social, cultural, and political impacts will only become apparent in the years ahead.

In Austria, the recent sociopolitical debate about the increasing privatization of beaches has been triggered both by Austria's highest rates in sealing soil per year in EU comparison and by socio-ecological critique of Austria's very high carbon footprint when quickly urbanizing land. Since 2022, a wider critical debate has been unfolding in which the nearly complete privatization of access to Austrian lakes has been criticized, with critics proclaiming the need for a political reregulation and repoliticization of planning processes to allow all people the constitutional access to public waters, both on public and private lands. A coalition claiming some kind of right to the beach was formed by a partnership between the Austrian Workers Chamber (Arbeiterkammer), the Friends of Nature (Naturfreunde), and the Alpine Association (Alpenverein) by starting from the premise that conflicts are increasing between the people seeking to enjoy nature as a commons and property owners.[47] On a regulatory level, these conflicts often result in collisions of constitutional rights;

for example, a basic right to free use of nature would be weighed against property holders' property right. These general reflections concerning the use of lakes have been disputed in Austria since the beginning of early industrialization. The partnership argues in favor of establishing a new constitutional right to enjoy nature, including access to natural lands, for the general public in Austria. While the interests of users and owners only rarely align, the report also confirms nature's (own) interest, paying attention to endangered species and protection of flora and fauna.

Acts of privatization of property at seasides and lakeshores and the restriction of public access to lakes can be understood as forms of institutional and political carelessness. They reflect a deep depoliticization of public affairs. Their repoliticization is important to secure the (potential or existing) constitutional right to enjoy nature and eco-cultural heritage through both the establishment of adequate legal and planning frameworks (Austria) and through their control and enactment in management and planning (Greece). In Greece, such a constitutional right is in place, yet it is not controlled and ratified in short-term management and long-term planning; in Austria, the claim to institutionalize such a right rises in a phase of striking depoliticization in processes of public neglect and structurally eroded provision of common lands, especially around lake access and beaches. Here, the introduction of new public management procedures regarding nationwide forestlands turned into the main facilitator of post-politicization. Despite the established constitutional right to access the beach, depoliticization in Greece happens through deceptive planning of beaches for private motives following a Greek phase of institutional restructuring induced by supranational entities.

Learning from the Post-Political Beach

We now move from empirical to analytical and theoretical lessons relating to the post-political beach. On the *empirical level*, disputes around the shrinkage—quantitative as well as qualitative—of beaches have been used to identify diverse impacts of neoliberal privatization. In the Austrian case, structural decisions taken in the 1990s privatized state-owned companies and turned state-owned common goods (under public law, with clear protection of its use value) into state-owned property (under private law, with clear promotion of exchange value). This led to the structural selling-off and beneath-the-radar building activity of lakeshores in which the state was the main agent triggering political transitions toward enhancing market dynamics around beaches. The Greek case also shows neoliberal institutional reforms in favor of market dynamics of the same genre initiated through imposed supranational operations. Additionally, the formulation of a concrete path of speculative urban development at the seaside is mainly driven by corporate consulting firms and (inter) national investors through governance beyond the state.

Both developments have been triggered by the EU and its policies for both national state governments (Austria in the 1990s, Greece in the 2010s) with the requirement of reducing state debts and budget deficits. In both countries, implicitly in Austria and explicitly in Greece, austerity constituted a *justifying mantra* to establish "a resurgent neo-liberalism with aims and practices which include disempowering and dismantling systems of social protection . . . , re-structuring, rescaling and downsizing the state . . . , and shifting the locus of risk and responsibility on to the public and to the poor in particular."[48] On the *methodological level*, we have identified connections between scales of neoliberalization, depoliticization, and post-politicization. Where studies of the neoliberal condition have accompanied the study of privatization of public spaces for decades, the scales of depolitization have seldom come under scrutiny. We identified specific beaches around which the extent of information seemed sufficient to also identify political-institutional scales at which local forms of depoliticized beach debates and development were triggered—the main drivers here were the EU and its allied supranational players, and the nation states. Yet, it is the local beach where the disruptive and unsettling impact of structural adjustments gets vividly manifested. This, however, is accompanied by a lack of civic engagement to fight for a civic right to use the beach as nature and as civic imaginary.

Are beaches then the emblematic symbol of consumer culture since more and more people have become acquainted with restricted access and luxury beach resorts, promoted through years of cheap beach tourism? As Hoskyns confirms, neoliberalism "strives to achieve a minimal state, which involves a dismantling of the public realm, and includes a transfer of public space to the private sector. These combined factors can be seen to have broken the links between democracy and public space, and produced a de-spatialization of democracy and a de-politicization of public space."[49] With regard to the beaches, institutional bodies of representative democracy have actively disregarded the institutionalization, execution, or control of the constitutional protection of the use of the beach as a common good, thereby actively ungrounding the principle of free beach access. Meanwhile, representative democratic bodies in Austria and Greece have purposefully decided to basically shift the character of beaches as a freely accessible public good. The accessibility of public space may serve as an indicator of quality of democratic life in a city, region, nation, or supranational realm.[50]

The other side of depoliticization concerns the erosion of the public sphere as the place where ideally public opinion around public affairs should be fiercely formed. Transferring Hoskyns's ideas further to the analysis of the beach as post-political, the depoliticization of the beach implies the disregard to beaches' capacities to serve as common goods. Common goods need to be democratically cared for by the people as well. The disregard of this principle has "weakened formal democratic practice in everyday life . . . [resulting] in a lack of articulation about what constitutes contemporary democratic practice and therefore the democratic spaces." In resonance with ideas of French philosopher Claude Lefort, Hoskyns coins the democratic public space as not occupied by one political position and, in that sense, as the empty space of democracy.[51] This concerns the role of the state but also all the instances designed to control state action, such as courts, universities, theaters, libraries, cinemas, and beaches, as they all have a role to play in nurturing critical counterpublics. The absence of these (spaces of) critical counterpublics may have different reasons: lack of transparency of processes of privatization and beach enclosure, lack of critical media in raising these issues as problematic, lack of resources and motivation to engage in political activism. The evident shift toward governance in planning implied a displacement of the public sphere so that issues of pressing public concern "become sheltered in shadowy forums comprised of select groups of influential stakeholders."[52]

On *the conceptual level*, we can follow Dikeç's method to identify space, and in this case, beaches, as a mode of political thinking. Transferring his insights to beaches, then, means that "any consideration of the relationship between space [or beach] and politics has to come to terms with this question of change. Whether understood as a sensible manifestation of things, an order of relations, a domain of experience, an analytical tool or as a mode of thinking, space [or beach] has to be associated . . . to change as a generative rupture in the order of things . . . to have political pertinence."[53] Hence, beaches cannot simply be explored "as a space through which politics is staged, but one in which the stage is constantly being constructed and reconstructed."[54] Acknowledging that politics is latent in all aspects of the beach and with all groups, "we must work towards creating . . . politics whereby all constituencies, and their grievances, have to equally answer to democracy's founding principle."[55] This means that "in all spaces and times there lies the potential for people to test existing social orders against the presumption of equality. The *method of equality* is offered here as a means to produce critical urban geographies that make a contribution to such enactments of democracy in the face of urban inequality."[56] Integrating Rancière's approach in urban studies helps as "his work seeks to address the question of what makes politics political."[57] The fact that the beaches analyzed have not been intensively politically reclaimed so far does not mean that there is not the potential to be reclaimed one day. The issue is to bring them back into Castoriadis's version of collective imaginary, in which a society self-institutes a different vision of beach which goes beyond its current post-political condition.

Conclusion

How democratic, how political, then, is the beach? Or is the beach in Austria and Greece rather a key object of state action and governance structures that can be coined as post-political? In this chapter, we have focused on transitions of beaches and the basic right of free use and access to beaches. By analyzing beaches with an emphasis on their (post-)political dimensions, we explored the hypothesis that they are an ambivalent arena where political acts to depoliticize public affairs become visible. As a result, we have found evidence that repeated sequences of depoliticization over time have accumulated into wider structural constraints around the governmentality of beaches which we have coined as post-political. In a few cases, struggles to repoliticize the beach have become apparent, yet a striking feature of the post-political beach is that people do not (yet) actively defend their *right to the beach*. However, to overcome the post-political state of things, "it is the political use of space . . . that does the most to reinstate use value."[58] *Beach* in this sense needs to become a space that encapsulates political thinking and reveals a terrain of contestation of representation and a spatiality of governmental orders. Finally, to develop a new collective imaginary, societies need not just to identify what they desire, but what they are lacking.[59] This chapter has shown that the narrative of the post-political beach is a story of a politics of abandonment of beaches as a common good, but also a story of inadequacy: lack of access to beaches, lack of protective governmental structures, and lack of people protesting and insisting to claim their own beach politics.

ACKNOWLEDGMENTS

We would like to thank Antonia Skenderović for her efforts to bring the manuscript into a final form, and Dragana Damjanovic for providing us with information as regards the legal situation of access to lakes in Austria. We are most thankful to Elizabeth Dickman for the language revision.

NOTES

1 Sabine Knierbein and Tihomir Viderman, *Public Space Unbound: Urban Emancipation and the Post-Political Condition* (New York: Routledge, 2018), 10.

2 Knierbein and Viderman, *Public Space Unbound*, referring to Jacques Rancière, *Dissensus: On Politics and Aesthetics*, ed. and trans. by Steven Corcoran (London: Continuum International, 2010), 9.

3 Erik Swyngedouw, "The Post-Political City," in *Urban Politics Now: Re-Imaging Democracy in the Neoliberal City*, Urban Politics Now Reflect Series, ed. BAVO (Rotterdam: Netherland Architecture Institute [NAi] Publishers, 2007).

4 Mustafa Dikeç, "Space as a Mode of Political Thinking," *Geoforum* 43, 4 (June 2012): 669–76.

5 Johanna Hoernig and Henrik Lebuhn, "Raumproduktionen: Inspirationen aus aktuellen kritischen Debatten; Ein Resümee," in *Raumproduktionen II: Theoretische Kontroversen und politische Auseinandersetzungen*, ed. Anne Vogelpohl, Boris Michel, Henrik Lebuhn, Johanna Hoernig, and Bernd Belina (Münster, Germany: Westfälisches Dampfboot, 2018), 178–88.

6 Jonathan Metzger, Philip Allmendinger, and Stjin Oosterlynck, *Planning against the Political: Democratic Deficits in European Territorial Governance* (New York: Routledge, 2015).

7 Japhy Wilson and Eric Swyngedouw, *The Post-Political and Its Discontents: Spaces of Depoliticisation, Specters of Radical Politics* (Edinburgh: Edinburgh University Press, 2015).

8 Jeffrey Hou and Sabine Knierbein, *City Unsilenced: Urban Resistance and Public Space in the Age of Shrinking Democracy* (London: Routledge, 2017).

9   Knierbein and Viderman, *Public Space Unbound.*

10 Knierbein and Viderman.

11 Metzger, Allmendinger, and Oosterlynck, *Planning against the Political*, 1–3.

12 Ali Madanipour, "Rethinking Public Space. Between Rhetoric and Reality," *UR-BAN DESIGN International* 24, no. 1 (2019): 38–46.

13 Dikeç, "Space as a Mode of Political Thinking," 669–76.

14 Dikeç, 670.

15 Cornelius Castoriadis, *The Imaginary Institution of Society* (Cambridge: Polity Press, 2005); Vangelis Papadimitropoulos, "Politics and the Political in Castoriadis," *Critical Horizons* 20, 1, (January 2019): 40–53, https://doi.org/10.1080/1440991 7.2019.1563996.

16 Carolina Crijns, *Architecture in Times of Multiple Crises. Everyday Utopianisms of Care and Radical Spatial Practices* (Bielefeld: Transcript, 2023).

17 Castoriadis, *Imaginary Institution of Society*, 160.

18 Castoriadis, 115.

19 Crijns, *Architecture in Times of Multiple Crises*, 21.

20 Cf. Crijns.

21 Dikeç, "Space as a Mode of Political Thinking," 673–74.

22 Teresa Hoskyns, *The Empty Place: Democracy and Public Space* (London: Routledge, 2014), 4.

23 Articles 5.1 and 24, Constitution of Greece (Government Gazette of the Hellenic Republic, No.187A, November 28, 2019).

24 Lazaros Karaliotas, "Geographies of Politics and the Police: Post-democratization, SYRIZA, and the Politics of the 'Greek Debt Crisis.'" *Environment and Planning C: Politics and Space* 39, no. 3 (2021): 491–511.

25 Erik Swyngedouw, "Post-democratic Cities: For Whom and for What?" Paper presented in concluding session at the Regional Studies Association Annual Conference, Pecs, Budapest, May 26, 2010.

26 City of Thessaloniki, *Resilient Thessaloniki: A Strategy for 2030* (Thessaloniki: Metropolitan Development Agency of Thessaloniki, 2017), 114–30.

27 World Bank Group, "Call for Expression of Interest: Waterfront Redevelopment Strategy for Thessaloniki #1246887," August 31, 2017.

28 Deloitte Business Solutions SA, *World Bank Group: Thessaloniki Waterfront Redevelopment Strategy. Framework Plan* (Thessaloniki: Deloitte Business Solutions SA, 2018).

29 Charis Christodoulou, "Public Space Shrinkage in the Current Liquidity of Processes in Urban planning and Development in Greece," [in Greek] *Geographies* (December 2020): 26–41.

30 Kontrast Redaktion [Kontrast editorial team], "Österreich, Land der Seen—doch in ihnen baden dürfen wir selten. Denn: Die Ufer gehören einigen wenigen," *Kontrast*, last modified July 14, 2021, https://kontrast.at.

31 Lisa Kreutzer, "Der Seehandel im Salzkammergut" [Lake Trade in Salzkammergut], *Falter*, 31 (2019): 2.

32 Kreutzer, 7.

33 Kreutzer.

34 "Privatisierung von Staatsvermögen: ÖIAG, Bundesimmobilien, Bundesforste," *Arbeit & Wirtschaft*, last modified March 14, 2001, www.arbeit-wirtschaft.at.

35 Peter Weinfurter, *80 Jahre Bundesforste: Geschichte der Österreichischen Bundesforste* (N.p.: Österreichische Bundesforste AG, n.d.), 71.

36 Weinfurter, 72.

37 Weinfurter.

38 "Interaktive Karte: Naturbadeplätze—Ungetrübtes Badevergnügen an den schönsten Seen," Österreichische Bundesforste, accessed May 22, 2024, www.bundesforste.at; "Bundesforste starten mit neu gestaltetem Naturbadeplatz am Attersee in die Sommersaison," Österreichische Bundesforste, accessed May 22, 2024, www.bundesforste.at.

39 Kontrast Redaktion [Kontrast editorial team], "So können alle von Österreichs Seen profitieren—nicht nur die Reichsten," *Kontrast*, August 4, 2022, https://kontrast.at.

40 "So können alle von Österreichs Seen profitieren."

41 Christine Steiner-Watzinger, "Attersee: Freie Seezugänge sind teils heiß umkämpft," *Mein Bezirk*, June 24, 2020, www.meinbezirk.at. This source also offers photos of the two luxury apartments and the public lido area.

42 Lukas Dierer, "Raumplanerischer Umgang mit Zweitwohnsitzen im oberösterreichischen Seengebiet Attersee und Traunsee. Problematik und Steuerungsmöglichkeiten," (Master's thesis, TU Wien, 2020).

43 Dierer, 92.

44 Dierer, 44.

45 Lazaros Karaliotas, "Geographies of Politics and the Police."

46 Evangelia Athanassiou, Charis Christodoulou, Matina Kapsali, and Maria Karagianni, "Hybridizing 'Ownership' of Public Space: Framings of Urban Emancipation in Crisis-Ridden Thessaloniki," in *Public Space Unbound: Urban Emancipation and the Post-political Condition*, ed. Sabine Knierbein and Tihomir Viderman (London: Routledge, 2018), 251–65.

47 Michael Ganner, Samantha Pechtl, Wolfgang Stock, and Karl Weber, "Recht auf Natur, Freier Zugang zur Natur," in *Informationen zur Umweltpolitik 204* (Vienna: Arbeiterkammer Wien, 2022).

48 Annette Hastings, Nick Bailey, Glen Bramley, and Maria Gannon, "Austerity Urbanism in England: The Regressive Redistribution of Local Government Services and the Impact on the Poor and Marginalized," *Environment and Planning A* 49, 9 (June 2017): 2007–24.

49 Hoskyns, *Empty Place*, 4.

50 Ali Madanipour, Sabine Knierbein and Aglaée Degros, "A Moment of Transformation," in: *Public Space and the Challenges of Urban Transformation in Europe*, ed. Ali Madanipour, Sabine Knierbein and Aglaée Degros (London/New York: Routledge, 2014), 1–8.

51 Hoskyns, *Empty Place*.
52 Metzger, *Planning against the Political*, 3.
53 Dikeç, "Space as a Mode of Political Thinking," 675.
54 Mark Davidson and Kurt Iveson, "Recovering the Politics of the City: From the 'Post-Political City' to a 'Method of Equality' for Critical Urban Geography," *Progress in Human Geography* 39, 5 (May 2014): 557, https://doi.org/10.1177/0309132514535284. Quoted by Sabine Knierbein, "Critique of Everyday Life in the 21st Century. Lived Space and Capitalist Urbanization," Unpublished Habilitation Treatise at TU Wien (2020), 367.
55 Davidson and Iveson.
56 Davidson and Iveson.
57 Davidson and Iveson, 547, referred to in Knierbein, "Critique of Everyday Life in the 21st Century," (2020), 367.
58 Henri Lefebvre, *The Production of Space*, 27th ed., trans. Donald Nicholson-Smith (Maiden, MA: Blackwell, 2009 [1974]), 356.
59 Castoriadis, *Imaginary Institution of Society*, 146.

PART II

Shoring Up the Coastline

Protection and Resistance

Wind from the West (Setha Low, 2023)

6

Beachfront Protection as Beach Privatization

Coastlines, Property Lines, and Climate Change
Adaptation in Sydney

KURT IVESON AND ANA VILA-CONCEJO

In June 2016, severe storms lashed Sydney's coastline. While many beaches along the coast were eroded significantly by these storms, the fate of one part of the coast came to dominate the news. Collaroy and Narrabeen, two adjoining beaches which form the 3.6-kilometer-long stretch of coast that is Sydney's most capitalized beachfront, suffered significant erosion during the storm. Ten beachfront properties were significantly damaged, and mainstream and social media circulated spectacular images of gardens, decks, furniture and even a swimming pool collapsing into the water (figure 6.1). How should the risks facing highly vulnerable beaches like Collaroy-Narrabeen be managed? As

Figure 6.1. Beachfront properties on Collaroy-Narrabeen Beach, June 2016. Courtesy of University of New South Wales Water Research Laboratory/Chris Drummond.

climate changes, sea levels will rise and severe storms like the one experienced in 2016 are expected to become more frequent. The way that public planning authorities, beachside communities and publics, and private property owners respond to these kinds of events will have profound consequences for public access to beaches in Sydney, and indeed to city beaches globally.

On urbanized coastlines, beaches have long been sites of conflict over questions of private and public ownership, access, and amenity. All public spaces are shaped by conflicts over the location and meaning of the property lines that demarcate public space as "the people's property" from private property.[1] However, the contested demarcation of public and private property lines and property rights in and around beaches takes on distinct characteristics because of the material dimensions of coastlines. Along with the daily movement of tidelines, beaches are subject to processes of erosion and accretion that take place over different timescales. As such, coastlines *move* and beaches "cannot be seen as fixed in time or space."[2] Questions of how property lines are drawn in relation to moving coastlines bear directly on the demarcation of spaces that are publicly accessible or exclusively private—does the public space of the beach move with the coast, or do private property lines remain fixed regardless of coastal processes?

The public-private tensions generated by coastal dynamics are likely to be heightened in the coming decades as climate changes. While beaches may need to migrate landwards in response to sea-level rise and changing wave climates, property owners typically feel the need to defend their property and property boundaries against any and all encroachments. Put bluntly, the public accessibility and amenity of beaches will be defined by whether the needs of the beach or of the property owners take priority. Our purpose in the chapter is to work through these relationships between coastlines and property lines in a changing climate and the significance of these relationships for the beach as a public space accessible to diverse urban publics.

We examine these broader questions through a case study of Collaroy-Narrabeen Beach, one of Sydney's northern beaches. The conflict over coastal protections that ensued after the damage to private property caused by the 2016 storms is emblematic of the kinds of public-private conflicts that will surely become more frequent. Years of fractious conflict involving property owners, beach communities and users, local and state government authorities and their paid consultants eventually resulted in the construction of a privately funded, state-approved seven-meter-tall seawall along a significant stretch of the beach. The wall is designed and justified as a measure to protect existing private properties and public infrastructure from further storm damage. The construction of this wall marks the emergence of a new, and troubling, approach to the management of beaches as public space in Sydney in the context of climate change. While this approach has the appearance of a democratically controlled coastal management process that brings the actions of beachfront private property owners under public control, in fact it privatizes beach management by placing the property regime itself beyond the reach of the policy process. The strip of sand that constitutes the beach, and by extension the beach's public and their access, is literally caught between a rising sea and more frequent storms on its seaward side and the rights of private property owners to maintain static property boundaries on its landward side. Here, we seem to be witnessing a kind of privatization of the public beach driven by the privileging of existing private property rights in the face of a retreating coastline.

To develop our argument, we begin with a brief orientation to beaches and beach formation, and then we survey the debates about climate change impacts and adaptation. From there, we consider how these dynamic beach morphologies interact with beachside urban development, focusing on the different forms that this has taken in Sydney. We then examine the case of Collaroy-Narrabeen's new seawall, situating it within the history of public-private relations in the development and management of that beach over the past century. As we show, the debates that took place in the wake of the 2016 storm are part of a much longer story of tensions between private and public interventions and interests in this beach. We conclude with a discussion about the ways in which private property is physically protected and politically entrenched in contemporary coastal management, and we speculate about the alternatives to this form of privatization. We argue that the protection of the beach as public space will require a break with current path dependencies that protect private beachside property, one which puts the *sand rights* of the beach and public rights of access ahead of private property rights.

Orientations: Coastlines, Property Lines and Climate Change

We begin with some "beach 101," via some basic coastal geomorphology which is vital to the development of our subsequent analysis. A beach is a wave-lain deposit that needs three factors to exist: unconsolidated sediment, waves to move it, and accommodation. Beaches extend within the limits of wave action within several meters above and below mean sea level.[3] Beach stability depends on the balance between erosion and accretion processes, that is, how well and quickly a beach recovers after an erosive event. Beach erosion is a well-known process on open-ocean beaches, and a plethora of numerical models can reproduce to a great extent erosive processes on open-coast beaches.[4] Beach recovery is a complex process in which the local availability of sediment is just one factor;[5] the balance between storm response, storm frequency, and recovery rates controls long-term coastal evolution. In particular, the worst coastal damage can occur due to storm clusters, when storms occur in close succession, not allowing for sufficient recovery time, which can produce years of coastal erosion within weeks or months.[6] However, our understanding of coastal storm response is limited by the quality and appropriateness of the data sets available, particularly for measurements throughout a full sequence of beach recovery.[7]

These underlying beach erosion and accretion processes are impacted by climate change, with its associated increases in sea level and storm frequency. Certainly, coastal erosion is not new—sandy coastlines around the world have been eroding as sea levels rise since the last ice age. But this process has recently begun an unprecedented acceleration. Through the record of the last 250 years, we can trace an increasing occurrence and magnitude of weather patterns and events that are now rendering beaches, and their associated habitats, vulnerable to forecasted dramatic changes caused by climate change.

While there is general agreement that climate change will impact sandy coastlines and beaches, the specifics of those impacts continue to be debated. International studies posed that almost half of the world's beaches would be gone by the end of the twenty-first century;[8] many coastal geomorphologists and engineers, however, argue that many beaches will successfully adapt to such changes.[9] In the Australian case, the debates about sea-level rise and coastal retreat are a case in point. In its 2021 report, the Intergovernmental Panel on Climate Change (IPCC) predicted with "high confidence" that "a majority of sandy coasts in the [Australasian] region will experience shoreline retreat, throughout the twenty-first century," with over ten thousand kilometers of coastline predicted to have retreated by more than one hundred meters by 2100 under the more dire global warming scenarios.[10] Others have argued that shoreline retreat cannot be modeled in the manner conducted by the IPCC authors. Andrew Short has argued that coastlines and coastal systems vary considerably and "can exhibit considerable spatial variation in behavior, even between adjoining beaches . . . and along the same beach." As a result, he argues that "predicting coastal behavior has to be undertaken at a local level."[11]

Short illustrates his claim about spatial variation in climate change impacts with reference to decades of data he has analyzed about beach dynamics and coastal erosion at Collaroy-Narrabeen. This beach, like many along Sydney's coast, is an embayment formed by headlands at either end which extend offshore as rock reefs. While storm events interacting with existing sea-level rise have certainly led to dramatic beach erosion episodes like the one in 2016, sand lost from the beach has not been lost from this beach embayment, which acts as a closed sediment compartment. The sand has always returned over time, and "sand volume in 2021 remains essentially the same as in 1976 when beach profiling commenced."[12]

Regardless of this ongoing debate, the general point holds that climate change will have diverse impacts on the natural dynamics of coastal erosion and accretion on which beach stability depends. While coastlines and coastal landscapes have never been "fixed in time and space" and "tidal boundaries are continuously moving,"[13] climate change will continue to disrupt and destabilize the boundaries between land and sea. In contexts where there is no human settlement or infrastructure, beaches can simply move in response. But in urban coastal contexts where beachfronts are developed, things are not so simple. As Angus Gordon puts it bluntly, there are ultimately two options for adaptation strategies where beach movement will impact on public and private assets: *managed retreat* which considers moving private property and public infrastructure to allow natural beach dynamics, or *defense of property* which typically occurs through engineered fortifications designed to prevent those dynamics.[14]

Variations in beach location and width associated with coastal and climate dynamics interact with urban dynamics in fraught ways. Capitalist urbanization is fundamentally underpinned by regimes of property that facilitate the ownership and exchange of defined parcels of land. However, in coastal urbanization, as Phil Hubbard notes, "the shoreline boasts specific material properties and capacities which constantly challenge legal attempts to fix patterns of land use and property."[15] Hannah Power and colleagues speak of an imagined "line in the sand" in our dominant imaginaries of beaches and beachfront property: "On one side of the line, you can buy land that is yours to use and develop until you sell it, with property boundaries remaining fixed regardless of any change to the landscape or movement of the coastline. But these embedded administrative constructs are incompatible with the reality of changing coasts as the effects of climate change are increasingly realized."[16] As the physical beach changes, what will take priority: the fixity of land use and property lines that are notionally established in perpetuity or the mobility of coastlines and tidelines that move over time?

Property regimes in different parts of the world have grappled with this question in different ways. In many jurisdictions, property lines at the beach are drawn in relation to tidelines. The question of where private property and associated rights begin and end depends on (1) laws and norms about public ownership of, and access to, beach sand seaward of the tideline and (2) whether or not beach sand and dunes that are landward of the tideline are preserved against all forms of development, reserved for public easements and access, or made available for private ownership and development. The movement of beaches subject to different property regimes ultimately challenges all of them, no matter their approach to these fundamental matters. Strategies will need to address the path dependencies already created by past decisions and will create their own path dependencies that will need to be dealt with by authorities and publics in the future.[17] As Hubbard argues, "The material forms of the coast, and its regulation have developed over decades via a three-way symbiosis in which the law has responded to environmental conflicts and problems but has also *produced* those conflicts."[18] The ongoing history of beachfront development through settler-colonial property regimes in Sydney since invasion are a case in point.

Coastal and Beach Management in Settler-Colonial Sydney

Sydney's early colonial settlement was along the foreshore of Port Jackson, the city's now-famous harbor. Coastal beaches to the east, north, and south that are lands of Eora, Garigal, and Dharawal peoples were initially relatively inaccessible to Sydney's colonizers. But as the city's population grew, and as transport infrastructures were gradually extended to beachfront areas, Aboriginal peoples were dispossessed of their country and colonial property was created, granted, and subdivided. Initially, some beaches were converted into private property; for example, the iconic Bondi Beach was private property for decades. But in most cases, private property rights did not extend beyond the foreshore on the ocean coastline. Gradually, in cases like Bondi, local governments wound back private property rights over the beach to provide for public access.

Across Sydney's beaches, early development took a range of forms. In some places, dunes and other natural landscapes adjacent to the beach were left in place, with property lines set back at a reasonable distance from beaches and their tidelines. In other places, public and private assets were developed much closer to the beach. Ocean pools were carved out of rocks at several beaches and seawalls were constructed to allow sand-free recreation on public promenade parks. Extreme cases of this enthusiasm for coastal recreation include the eastern beaches of Coogee and Tamarama, where a fun pier and an amusement park with a rollercoaster had a short-lived existence. Beach and surf club facilities are prominent alongside private properties developed on beachfronts.

In the early days of beach development, any private property tended to be occupied by basic accommodations often used for fishing and beachside holidays. But as car ownership and mass transit developed, and distances to the beaches effectively shrank, properties started to be redeveloped into denser and grander accommodations, and prices rose to match.[19] Most beaches are now located in exclusive real estate markets, and economists estimate that being on the beachfront itself adds a 40 percent premium in comparison to other adjacent, nonbeachfront properties.[20]

However, while most Sydney-siders cannot afford to live close to a beach, public access to these iconic public spaces is jealously guarded.

Beaches are among Sydney's most iconic public spaces, all open-coast beaches are publicly accessible,[21] and most are home to community-run surf lifesaving clubs and facilities. Swimming, surfing, suntanning, and socializing at the beach are important parts of Sydney's cultural fabric. A recent proposal to enclose part of iconic Bondi beach for commercial seating that would be available for beach patrons to rent, with service of food and drinks, was widely condemned and ultimately rejected.

Sydney's coastline and beaches have proven to be hazardous areas for urban development. Almost as soon as public and private property owners began to build near beaches, some of the structures they built were damaged or destroyed during major weather events that occurred every decade or so. Perhaps most famously, the Coogee Pier that was opened in 1928 was short-lived; rough surf damaged the pier, and it was demolished in 1934. But beyond this high-profile example, surf clubs, seawalls, ocean pools, promenades, and private properties have all been subject to the harsh conditions that frequently pertain at the coast.

As of now, even without the complicating factor of climate change, many of the aging public and private properties and assets on beach-fronts are posing challenges for coastal planning and management. As climate changes, and as storm events and storm clusters which cause short- and long-term erosion of beaches occur at greater frequency and from changing directions, such challenges are becoming acute in many areas. While the understanding of coastal processes and engineering have advanced considerably since these initial developments, the kinds of path dependencies of existing planning and property regimes discussed in the previous section mean that these advances are never applied to the beach as a "clean slate."

For most of the twentieth century, management of such issues was typically left to local governments, which do not necessarily command the powers or resources to address coastal hazards and vulnerability along entire beaches. As a consequence, while local governments have acted to protect their own assets, there is a long history in Sydney and elsewhere on Australia's east coast of private property owners "defending their assets during storms by unauthorized, ad hoc, tipping of rock and other materials in attempts to stop the loss of their assets."[22] These private works are generally recognized to have had negative impacts on neighboring properties and beach amenity.[23] They impinge on the publicness of public beaches in two senses: by reducing the size of the physical public space and by asserting private rather than public control over the management of the beach.

In an effort to introduce more consistent and comprehensive public control over coastal development and management, the New South Wales (NSW) government first introduced the NSW Coastal Protection Act in 1979.[24] This was replaced with the NSW Coastal Management Act in 2016. These acts set the parameters for beachside local governments to develop their own Coastal Zone Management Plans, which in theory direct the action of public authorities and regulate the action of private property owners. In the next section, we examine the operation of the current regulatory and planning regime in Sydney's most vulnerable and most debated beach: Collaroy-Narrabeen.

Lines in the Sand: The Collaroy-Narrabeen Beach Seawall

Collaroy-Narrabeen is one of Sydney's most iconic surfing beaches. It even appears in the lyrics of the Beach Boys' hit "Surfin' U.S.A.," among its many claims to fame. This beach has a long history of significant erosion events, which have impacted the public beach and beachfront properties. The first land grants in the area were made in the early 1800s. This land was initially subdivided in the early 1900s, and the subdivision at the beachfront involved the removal of the beach dune cap and associated vegetation.[25] As on many beaches in Sydney, the first settler-colonial structures were community facilities like surf clubs and dressing sheds as well as small, privately owned beachfront shacks. Not long after, in 1913, Collaroy Surf Club and the dressing sheds were destroyed by beach erosion caused by significant storm events. A little over a decade later, in 1925, five beachfront houses were destroyed and a further ten were significantly damaged.[26] Those properties were then purchased by the local government and converted into public open space, and in the 1930s, a seawall was constructed at the southern end of the beach in front of redeveloped beach club facilities. More property buybacks and conversions occurred again after severe storms destroyed more properties in the mid-1940s.

The pace of development picked up significantly in the 1950s with the growth of car ownership and the construction of new roads from the city; this included the first apartment buildings on the beachfront. After more storm events in the 1960s, the local government authority initiated a policy of purchasing at-risk properties when they came on the market for conversion into public space. This policy continued into the early 2000s, but by this point, it had become increasingly difficult for the local government to continue, given soaring prices for beachfront properties. By the time of a 2005 purchase, it cost close to AUS$3 million (approximately US$1.9 million) for a detached house, making buybacks and conversions unsustainable.[27]

Until this point, these local government interventions were a reactive strategy of coastal retreat, which did have the effect of buffering a substantial proportion of the beach with restored dunes and parkland. Nevertheless, despite almost a century of efforts to leverage state resources to give the public beach more room to move, Collaroy-Narrabeen is now Sydney's most capitalized beachfront.[28] For their part, beachfront private property owners were also active in beach management at Collaroy-Narrabeen, installing their own property protections on the seaward side of their properties. These defensive structures were frequently unapproved, uncoordinated, and extended beyond the private property boundary onto the public beach. As well as impinging on the public beach space, such interventions can have negative effects on beach stability. The local government unsuccessfully sought to have some of those unauthorized structures removed; one property owner won a case in the Land and Environment Court to prevent the removal of an unauthorized seawall they had built to protect their home in 1997, following a storm at that time.[29]

Shifting its strategy, in the early 2000s, the local government began to investigate the construction of an engineered seawall to replace the existing protections installed by property owners. However, the impacts and ideology of engineering interventions in the face of natural hazards faced more critical scrutiny thanks to a combination of decades of coastal science and a growing environmentalist movement which questioned human efforts to tame nature through development. The local branch of Surfrider Foundation led a campaign against the proposed seawall, including a "Line in the Sand" protest, which involved around three thousand people forming a kilometer-long human wall along the beach.[30] Campaigners asserted that while the proposed wall might protect properties on the beachfront, it would harm the beach. According to one of the organizers, "Seawalls do nothing to ensure the ongoing conservation of the beach in front of them. Worldwide experience shows us that they actively destroy it."[31]

In the face of this campaign, disagreements among property owners, and debates about who would pay, the seawall proposal was abandoned. In retrospect, we can see that this conflict marked the beginning of a period in which management strategies began to shift from managing coastal retreat via property acquisition and dune restoration toward defending beachfront property which has become increasingly valuable.

A long-term management plan that covers both the beach and lagoon behind it has been implemented since the mid-1970s, with two main elements.[32] First, the dredging of the flood-tide delta of the entrance to the lagoon (Narrabeen Lake) minimized flooding risks for the properties and infrastructure surrounding the lake. Second, the sandy sediments dredged were then transported by truck to the vulnerable Collaroy beach, thus replenishing the beach and decreasing its vulnerability to coastal erosion. This is a soft engineering intervention that still allows natural processes to occur; the natural southeast waves that dominate the NSW wave climate then slowly move that sediment back toward the north, where they eventually enter the inlet to form the flood-tide delta again. Meanwhile, the local government also sought to remove some of the unauthorized private protections that had been installed by property owners.

Despite these management interventions, the June 2016 storm highlighted the ongoing vulnerability of this beach to severe coastal erosion during storm events. Images of the very eroded beach, with property destroyed and even a swimming pool laying on the beach, inundated the media. Collaroy-Narrabeen became the poster beach for coastal erosion. While many other beaches in Sydney lost a larger percentage of their sand,[33] none of them received as much media attention and political debate. The June 2016 storm occurred during spring high tides with maximum wave heights propagating from a direction approximately 45 degrees more counterclockwise than average.[34] Typical storms in Sydney are defined as significant wave heights larger than three meters and have waves propagating from the southeast; in consequence, Collaroy beach is often protected from storm waves. The atypical direction of the June

2016 storm meant that vulnerable Collaroy received the full impact of the storm waves, causing the largest erosion since 1976.[35]

Not surprisingly, the adequacy of Collaroy-Narrabeen's management strategy was vigorously debated in the wake of the 2016 storm damage. Those who had been in favor of the 2002 seawall, and their supporters in the media, blamed the environmentalists who had campaigned against it and once again began pushing for a major engineering solution. Local government faced pressure to approve the construction of a new seawall and initiated a review of the existing Coastal Zone Management Plan for Collaroy-Narrabeen Beach. This review was hastily completed—a draft of the revised plan was placed on public display a little over a month after the June 2016 storms, and the final revised plan was published and adopted in December 2016. The stated objective of the Plan was to "find a balanced and achievable approach that protects and preserves beach environments while limiting the impact of coastal processes on public and private assets." It set out four key priorities:

1. Protect and preserve beach environments and beach amenity;
2. Manage current and projected future risks from coastal hazards;
3. Ensuring continuing and undiminished public access to beaches, headlands and waterways;
4. Protect or promote the culture and heritage of both beaches.[36]

This plan explicitly rejected planned retreat as an option, not on the grounds that it would not work, but rather on the grounds that it would be too expensive. The rejection of this approach is dispensed with in a single short paragraph: "Planned retreat was not considered feasible at Collaroy-Narrabeen Beach and Fisherman's Beach due primarily to the level of existing development and to the small lot size of the properties. Additionally, while property purchase has been undertaken by Council in the past it is now considered to be a cost-prohibitive management option for Council and not a priority for public expenditure."[37] As well as ongoing beach nourishment and dune management, the plan specified parts of the beachfront where new protection works such as seawalls would be suitable. In doing so, it made clear that the construction and maintenance of any works designed to protect beachfront property would be the financial responsibility of property owners.

The local government also commissioned a team of engineers to deliver a concept design for new seawall protections. The concept design modeled the removal of existing unapproved and uncoordinated protections and their replacement with protections that would be compliant with the Coastal Zone Management Plan and other relevant legislation. The engineers' report argued that research about the negative impacts of seawalls on beach erosion and coastal processes—both on the beach in front of walls and on sections of the beach adjacent to the walls—was "somewhat unresolved" with "consensus not obtained."[38] It found that the installation of a 1.7 to 2.0 kilommeter seawall, mostly located on private property and potentially in several construction stages, would reduce the risk of inundation and erosion to protected properties while having "no discernible adverse impacts . . . on coastal processes or amenity values."[39] Indeed, they asserted that a new seawall would improve the visual amenity for beachgoers "due to a more regular structure and less small material and detritus" and would not reduce the width of the beach by more than 2 meters in any area.[40]

Notably, however, such calculations were "based on a comparison with the current foreshore state, inclusive of the existing ad-hoc protection works, their present impacts (which have existed for several decades) and ongoing sand management practices."[41] So, in a neat maneuver, the rejection of planned retreat in the Coastal Zone Management Plan meant that impacts of new seawall protections were never compared with a planned retreat through beach and dune restoration along the specified part of the beachfront or with any other form of adaptation that might impinge on private property. While the report acknowledged that "the main cause of the existing coastal hazards is that development has taken place well within the active coastal zone," the challenge was framed as one of protecting that very development rather than including its ongoing viability as a matter for consideration.[42] This worked to crowd other priorities out of the analysis; for instance, North Narrabeen was declared a National Surfing Reserve in 2009 for the quality of its break,[43] but protecting or enhancing this quality of the beach is given no serious attention in the report.

Taking their cues from these local government policy changes and reports, a group of beachfront residents formed a corporation, commissioned an engineering firm to develop plans for a seawall that would be compliant with the revised Coastal Zone Management Plan, and submitted a development application. Despite ongoing objections and protests against the construction of the seawall on the grounds that it would ultimately harm the beach, it was approved by the local government in 2020, and the first sections of the wall were constructed in 2021. While most of the wall is located within the private property boundaries, it certainly constitutes a very visible modification to the public beach (figures 6.2 and 6.3). Taking for granted existing private property as *both* the cause of the hazard *and* the immovable object of protection is a textbook example of the path dependency that Tayanah O'Donnell argues has profoundly delimited the development of beach and coastal management plans. She argues that in the face of climate change and its associated impacts on beaches, "implicit acceptance of and assumptions about colonial property rights have directed the flow of adaptation policy pathways," embedding a "path dependency for coastal climate adaptation" informed by a "deferential prioritization of property and property rights."[44] This prioritization tends to push management in the face of climate change toward strategies like the Collaroy-Narrabeen seawall, in which "manipulation of coastlines via engineering responses [are] viewed as the preferred way to balance competing public and private interests in coastal locations."[45]

In fixing existing private property lines in place for years and decades to come through approval of the wall, the local government has potentially solved one problem by creating another. The local government, acknowledging that in the context of sea-level rise, and with a wall in place, the width of the publicly accessible beach may now shrink, had a strategy for maintaining the size of the beach: to secure the resources to add more sand to the beach compartment; "Sea level rise may result in progressive loss of beach width over coming decades. Council will work with the State Government and other coastal councils in NSW to facilitate the importation of sufficient quantities of sand to enable beach width and surf quality to be maintained."[46] The cost of this has not yet been calculated, nor has the partnership with other scales of government even been established. Caught between the cost of buying back expensive beachfront properties now and the cost of sand replenishment at some point in the future, today's local government has chosen the latter, kicking the problem down the road. And of course, as property values

Figure 6.2. A section of the completed Collaroy seawall.
Photo by Kurt Iveson.

Figure 6.3. Erosion during the Queen's birthday storm of
2007, at the location of the future seawall. Photo by Ana
Vila-Concejo.

are protected and enhanced through expensive protections, any change
of direction toward planned retreat in the future through property purchases will be even more expensive.

While the effects that the seawall will have on the beach remain uncertain, there is an element of certainty that it will end up affecting the
beach negatively. A vertical seawall has minimum wave dissipation ca-
pability and thus reflects the wave energy back to sea and to the adjacent
parts of the beach that, in consequence, may receive enhanced wave energy.[47] In other parts of the world, seawall construction has led to other

seawalls being constructed on the adjacent beach sections. This process can continue until the entire beach has a seawall at its back and may lead to beach disappearance. This is the fear of coastal geoscientist (and our colleague) Andrew Short, who worries that "eventually you'll end up with a seawall and no beach at all. What the council is doing is saying we prefer to protect a handful of beachfront properties rather than a very popular Sydney public beach."[48]

Conclusion: Coastal Adaptation and Beach Privatization—Sand Rights, Public Rights, and Property Rights

Even in the most optimistic climate change mitigation scenarios, sea-level rise, changes in wave direction, and increasingly frequent storm events are already locked into global climate. In those urban coastal contexts where rising seas and more frequent and intense storms (and their associated waves) will cause beach erosion or recession, or both, the implications for public access are troubling. As urban and coastal authorities make decisions about the relative priority given to property defense and coastal retreat in adaptation strategies, will public access be a priority that guides these strategies? On highly capitalized urban beachfronts, public access has always existed in a fraught relationship with private property ownership. There public-private conflicts are not simply conflicts over whether the beach is publicly or privately owned; even when the beach is notionally public property, the rights and actions of beachfront private property owners can shape access and atmospheres on beaches in all manner of ways.

The story of the Collaroy-Narrabeen is a cautionary tale about the privileging of private property in beach adaptation strategies, which has led to engineered protections that have harmful impacts on public access to, and public control over, the beach. Those responsible for proposing, supporting, approving, and constructing the seawall at Collaroy-Narrabeen have argued that it has no adverse impacts to public accessibility or amenity of the beach. And, on one level, they are right. The wall is mostly paid for by the private property owners who will see the most benefit, with local and state governments only contributing 10 percent each to the cost of the wall. The wall does not cut off access points to the beach any more than previous structures that it replaced. It currently only covers a relatively small proportion of the beach, and there are still many points of public access to the beach. Indeed, the seawall's supporters and approvers even make the claim that the wall has made the beach *more* public. Previously existing protections built by private property owners with no public discussion or state approval, including some which trespassed beyond private property lines onto the public beach, have now been replaced with a protective structure that has been subject to a planning process and accountable to policies enacted by democratically elected local and state governments. In this sense, a seawall designed to hold back storm surges and waves might be said to be quite different from walls designed to restrict access to premium privatized public spaces in other urban contexts such as gated communities—exclusionary walls which have been so contentious in the critical literature on contemporary public space.[49]

However, we would argue that the publicness of Collaroy-Narrabeen Beach is being significantly compromised by what O'Donnell has described as the "deferential" treatment of private property in the development of adaptation strategies. Beachfront property owners have capitalized on their proximity to and views of the public beach. The escalating premium that is attached to beachfront real estate is a private appropriation of value from nature and from public and community care for the beach itself. And in the coastal management process which led to the modeling and approval of a seawall, those properties and their appropriated property values are now beyond the reach of public action and debate. Instead, the public and supposedly democratic process for determining the best option for the beach's future are confined to discussing the best options for *protecting* those property values, while also maintaining public access to the beach. In this case, while the wall may ultimately result in the shrinking of the beach as a site of public sociability through accelerated erosion, it is also the product of the shrinking scope of public debates about the fate of the beach. The status of beachfront private property was literally put beyond question.

Even from a cost-benefit perspective, it may be cheaper to buy back properties at market prices now than to import sand for beach nourishment in an unspecified future. But, as things currently stand in Australia, commonwealth and state governments that have the resources for this tend to reserve their hazard-related spending to *reactive* responses to disasters that have already occurred.[50] There's no state or commonwealth fund that might be accessed by local governments for *proactive* coastal management involving property buybacks. And there's certainly no taste for either allowing those property values to decline through further exposure to damaging storms or for compulsory acquisition below market values. It's hard for us not to agree with Gordon's assessment that there's folly in thinking that engineering solutions such as the seawall and sand replenishment can hold back the coastal dynamics unleashed by climate change. For him, "It's a nineteenth century response to a twenty-first century problem. A step back in time to the non-environmental brutalist engineering solutions of the 1900s."[51] For Orrin H. Pilkey and J. Andrew G. Cooper, this logic is a threat to beaches everywhere: "The greatest threat to the future of the world's beaches is posed by the coastal-engineering profession. The engineer's primary charges are to protect beachfront buildings and enable navigation in and out of ports. To accomplish this, engineers attempt to hold shorelines still, despite the fact that flexibility is essential to the survival of beaches."[52] Ultimately, the only alternative is for the sand rights of the beach to move and the public accessibility of that sand to take priority over property rights of beachfront property owners.[53] For this to occur, we will need to wrest back public control over the fate of our public beaches rather than deferring to historical colonial constructs of private property.

NOTES

1 Lynn A. Staeheli and Don Mitchell, *The People's Property? Power, Politics and the Public* (New York: Routledge, 2008).

2 Bruce Thom, "Climate Change, Coastal Hazards and Public Trust Doctrine," *Macquarie Journal of International and Comparative Environmental Law* 8, no. 2 (2012): 28.

3 Andrew D. Short, *Handbook of Beach and Shoreface Morphodynamics* (New York: John Wiley, 1999), 379.

4 Óscar Ferreira, "The Role of Storm Groups in the Erosion of Sandy Coasts," *Earth Surface Processes and Landforms* 31, no. 8 (2006): 1058–60; Giovanni Coco, Nadia Senechal, Antoine Rejas, Karin R. Bryan, Sylvain Capo, Parisot Jean Paul, Jenna Brown. Jamie H. Macmahan, "Beach Response to a Sequence of Extreme Storms," *Geomorphology* 204 (2014): 493–501.

5 Nadia Senechal, Jonathan Pavon, Remy Asselot, Mohammed Taaouati, Sophie Ferreira, Stéphan Bujan, "Recovery Assessment of Two Nearby Sandy Beaches with Contrasting Anthropogenic and Sediment Supply Settings," *Journal of Coastal Research* 75 (2016): 462–66.

6 Ferreira, "Role of Storm Groups in the Erosion of Sandy Coasts"; Coco et al., "Beach Response to a Sequence of Extreme Storms."

7 Coco et al., "Beach Response to a Sequence of Extreme Storms."

8 Michael I. Vousdoukas, Roshanka Ranasinghe, Lorenzo Mentaschi, Theocharis A. Plomaritis, Panagiotis Athanasiou, Arjen Luijendijk and Luc Feyen, "Sandy Coastlines Under Threat of Erosion," *Nature Climate Change* 10 (2020), 260–63.

9 J. A. G. Cooper, G. Masselink, G. Coco, A. D. Short, B. Castelle, K. Rogers, E. Anthony, A. N. Green, J. T. Kelley, O. H. Pilkey, D. W. T. Jackson, "Sandy Beaches Can Survive Sea-level Rise," *Nature Climate Change* 10 (2020), 993–95.

10 Roshanka Ranasinghe, Alex C. Ruane, and Robert Vautard, "Climate Change Information for Regional Impact and for Risk Assessment," in *Climate Change 2021: The Physical Science Basis; Contribution of Working Group I to the Sixth Assessment Report of the Intergovernmental Panel on Climate Change*, ed. Valérie Masson-Delmotte, et al. (Cambridge: Cambridge University Press): 1767–926.

11 Andrew D. Short, "Australian Beach Systems: Are They at Risk to Climate Change?" *Ocean and Coastal Management* 224 (June 2022): 1.

12 Short, "Australian Beach Systems," 3.

13 Thom, "Climate Change, Coastal Hazards and Public Trust Doctrine," 28.

14 Angus D. Gordon, "The Failure of NSW Coastal Management Reform," *Coast and Shore* 89, no. 3 (2021): 50.

15 Phil Hubbard, "Legal Pluralism at the Beach: Public Access, Land Use, and the Struggle for the 'Coastal Commons,'" *Area* 52, no. 2 (2020): 421.

16 Hannah Power, Michael Kinsela, Thomas Murray, and Andrew Pomeroy, "Life on the Edge: Adapting Coastal Management in a Changing Climate," *Australian Quarterly* 94, vol. 93, no. 3 (July–September 2022): 12–20. https://aips.net.au/aq-magazine.

17 Tayanan O'Donnell and Bruce Thom, (2022) "Coasts: A Battleground in Disaster Prepardness, Response and Climate Change Adaptation," in *Complex Disasters, Resilience, Reconstruction and Recovery*, ed. A. Lukasiewicz and T. O'Donnell (Singapore: Springer, 2022): 81–97; Thom "Climate Change, Coastal Hazards and Public Trust Doctrine"; Gordon, "Failure of NSW Coastal Management Reform."

18 Hubbard, "Legal Pluralism at the Beach," 426.

19 Gordon, "Failure of NSW Coastal Management Reform"; Power et al., "Life on the Edge."

20 Griffith Centre for Coastal Management, *Assessment and Decision Frameworks for Seawall Structures: Appendix C Economic Considerations* (Sydney: Sydney Coastal Councils Group, 2013), 7.

21 This contrasts with many harbor and estuary beaches in Sydney, which do not have public access.

22 Gordon, "Failure of NSW Coastal Management Reform," 46.

23 Gordon, "Failure of NSW Coastal Management Reform"; Thom, "Climate Change, Coastal Hazards and Public Trust Doctrine."

24 There are over thirty local governments across Sydney's metropolitan area, but their planning powers are significantly shaped by the New South Wales govern-

ment, which is responsible for legislating planning regimes as well as environmental controls.

25 Warringah Council, "Appendix A: General Description and Photographs of Collaroy-Narrabeen Beach and Fishermans Beach," in *Collaroy-Narrabeen Coastal Zone Management Plan* (Manly: Warringah Council, 2014); Thom, "Climate Change, Coastal Hazards and Public Trust Doctrine."

26 Gordon, "Failure of NSW Coastal Management Reform."

27 Jack Houghton, "Sydney Storm: Massive 2002 Protest Stopped Sea Wall Being Built, but at What Cost?," *Daily Telegraph*, June 9, 2016, www.dailytelegraph.com.au.

28 Northern Beaches Council, *Collaroy-Narrabeen Coastal Protection Assessment* (Manly: Northern Beaches Council, 2016), 25.

29 Houghton, "Sydney Storm."

30 Thom, "Climate Change, Coastal Hazards and Public Trust Doctrine."

31 D. Smith and C. O'Rourke, "Wall of Humanity Lines up Against Councils," *Sydney Morning Herald*, November 18, 2002, 8.

32 D. W. Cameron, B. D. Morris, L. Collier, T. Mackenzie, "Management and Monitoring of an ICOLL Entrance Clearance," *16th NSW Coastal Conference*, accessed May 1, 2024, www.coastalconference.com.

33 Shari L. Gallop, Ana Vila-Concejo, Thomas E. Fellowes, Mitchell D. Harley, Maryam Rahbani, John L. Largier, "Wave Direction Shift Triggered Severe Erosion of Beaches in Estuaries and Bays with Limited Post-Storm Recovery," *Earth Surface Processes and Landforms* 45, no. 15 (2020): 3854–68.

34 Mitchell D. Harley, Ian L. Turner, Michael A. Kinsela, Jason H. Middleton, Peter J. Mumford, Kristen D. Splinter, Matthew S. Phillips, Joshua A. Simmons, David J. Hanslow, Andrew D. Short, "Extreme Coastal Erosion Enhanced by Anomalous Extratropical Storm Wave Direction," *Scientific Reports* 7 (2017): 6033.

35 Harley et al.

36 Northern Beaches Council, *Collaroy-Narrabeen Coastal Protection Assessment* (2016): i.

37 Northern Beaches Council, 50.

38 Northern Beaches Council, 25.

39 Northern Beaches Council, v.

40 Northern Beaches Council, 35.

41 Northern Beaches Council, 24.

42 Northern Beaches Council, ii.

43 See "North Narrabeen Surf," Surfing Reserves, accessed May 1, 2024, www.surfingreserves.org/north-narrabeen.

44 Tayanah O'Donnell, "Interrogating Private Property Rights and Path Dependencies for Coastal Retreat," *Ocean and Coastal Management* 231 (January 2023): 1, 2.

45 O'Donnell, 1.

46 Northern Beaches Council, *Northern Beaches Coastal Erosion Policy* (Manly: Northern Beaches Council, 2016), 5.

47 J. Andrew G. Cooper and Orrin H. Pilkey, eds., *Pitfalls of Shoreline Stabilization: Selected Case Studies, Coastal Research Library* (Dordrecht: Springer Netherlands, 2012).

48 Quoted in Freya Noble, "Huge $25 Million Concrete Wall Designed to Stop Coastal Erosion Could Wipe Out Beach Completely," *9News Online*, October 8, 2021, www.9news.com.au.

49 Teresa Caldiera, *City of Walls: Crime, Segregation, and Citizenship in São Paulo* (Berkeley: University of California Press, 2000).

50 Power et al., "Life on the Edge"; Gordon, "Failure of NSW Coastal Management Reform," 51.

51 Quoted in Wendy Harmer, "As Australia Battles Wild Weather and Coastal Erosion, We Should Learn from our Mistakes," *The Guardian*, 23 April 23, 2022. www.theguardian.com.

52 Orrin H. Pilkey and J. Andrew G. Cooper, *The Last Beach* (Durham, NC: Duke University Press, 2014), 41.

53 On "sand rights," see Thom, "Climate Change, Coastal Hazards and Public Trust Doctrine." See also O'Donnell, "Interrogating Private Property Rights."

7

Protection of the Shoreline

Tensions and Conflicts on Fire Island

DANA TAPLIN

This chapter explores the tensions that play out in how people and institutions seek to protect the environment as both a set of ecologies and as an inhabited social landscape. The site is Fire Island, one of ten National Seashores in the United States National Park System. Fire Island—historically the Great South Beach—is a long barrier island that forms Great South Bay along the south side of Long Island, New York (figure 7.1). Approximately eighteen seasonal communities exist on the island as inholdings within the federal property, protected by the legislation that created the National Seashore in 1964. As a coastal environment incorporated into a National Seashore, the National Park Service is charged with protecting the barrier island ecosystem, making it accessible to visitors, and interpreting its significance. Seasonal residents are protective too—fiercely so, as one resident said—not as much of the natural system as of the property and privilege they enjoy as inhabitants of an extraordinary natural environment. Seasonal migration of sand alongshore, dune erosion, and breaches in the island due to storms are all part of the natural cycle of an unstable coastal environment. The National Seashore is concerned with protecting those dynamic natural cycles that continually shape and reshape the island's contours, yet residents and property owners seek to arrest these forces to stabilize—make permanent—the land under their houses.

The conflicts over governance in a contested space reflect different meanings of protection: protecting an environment for dwelling, with its requirements of stable ground and property, and protecting the dynamic rhythms of the coastal environment. In the conflict over protection, I see a difference in acquiring knowledge through expertise and empirical

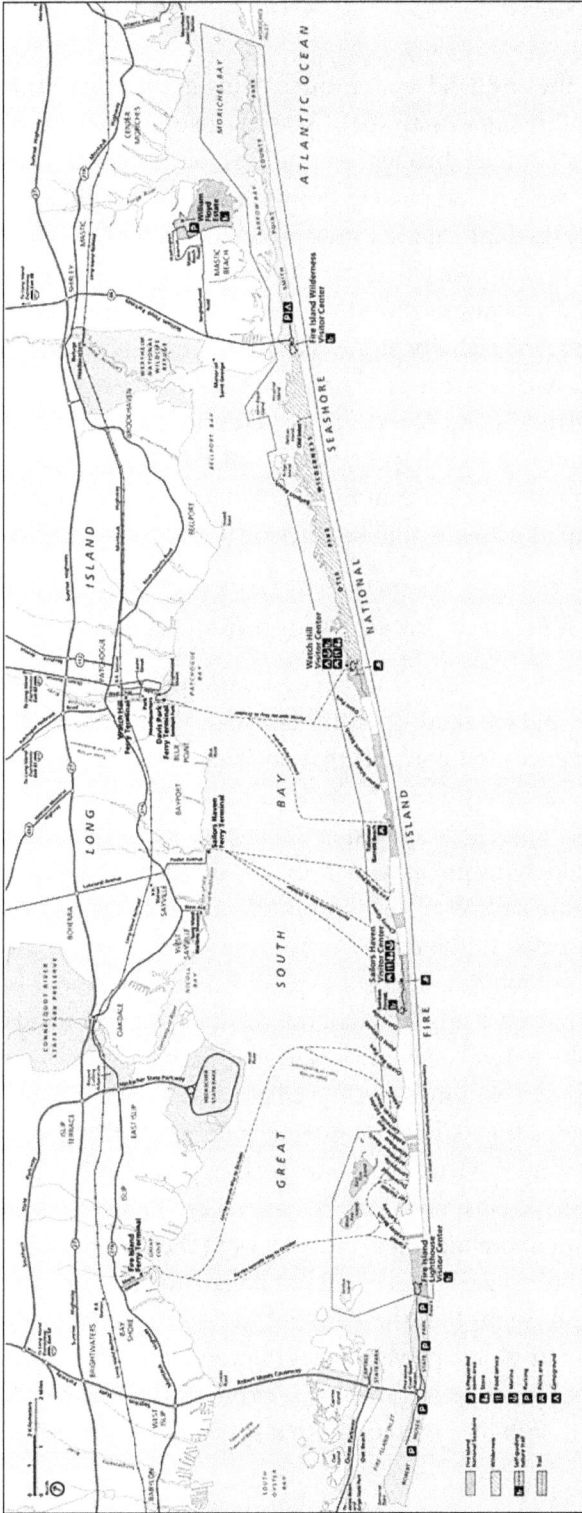

Figure 7.1. Area map showing the main sites of Fire Island National Seashore. Source: https://www.nps.gov/fiis/planyourvisit/maps.htm.

evidence, on the one hand, and through living in the place, fishing, boating, clamming, being witness to the seasons, telling stories, and handing down understandings through generations, on the other. On the Park Service side, we have employees doing their jobs to fulfill the institutional mission and taking their own protective initiatives. On the community side, we have people telling stories of community life and of the old days and displaying a readiness to explain what the authorities need to do to restore the stability of the island environment. Other actors are the over-and-back visitors—"day-trippers"—both to the National Seashore's public areas and to the various communities, especially those offering restaurants and shops to round out a visitor's day at the beach. There is also the business community, which is much more solicitous of the day trip and boating public than are the homeowning residents.[1]

What Does It Mean to Protect and Interpret the Seashore?

"Fire Island has this beautiful, striated dark sand, this gorgeous garnet, dark red purple sand. It's different than the white sand, you can feel the difference. You don't get that on Cape Cod. So you would bury that with ocean-dredged sand?" With these words, an urban planner at Fire Island National Seashore expressed her appreciation of one feature of the natural environment, which the seashore, informally called the "park," is authorized to protect and interpret for the public. Her challenge was directed to the more vociferous homeowners of the various Fire Island communities, who, she explained, twist the barrier island fragility question from one of inherent fragility to a fragility of indifference, as if to say the federal agencies—National Park Service, Army Corps of Engineers—are willing to let the island wash away for lack of care; that the island is fragile because no one is doing anything to protect it. "That's their war cry: 'Don't you want to save your park? You don't care about the park; you just want to let it wash away!' Rather than admit that building on the shoreline is unwise, they want the Park Service to maintain the line, spend gazillions of federal dollars to build up the beach."

It is beyond doubt that the Park Service cares about the beautiful, striated sand, the piping plovers, the waters and uplands and dunes, and the rest of the island environment that makes up its National Seashore. The conflict is with the "inholders," that is, the residents of the eighteen communities that precede the establishment of the National Seashore and who variously appreciate, resent, and upbraid the Park Service for interfering too much with their ways of life and not doing enough to preserve the balance of natural forces in a way that protects their houses and communities. For years, the communities have wanted the federal agencies to close breaches in the island and dredge sand to build up the beaches and dunes.

Settlement History

As Andrew Kahrl has observed, coastal domicile in the United States has increasingly shifted from subsistence communities and small-scale farmers, chiefly Black and Indigenous, to affluent, predominantly white residential communities who prize the recreational and restorative amenities of coastal residence.[2] With the transition—and to an extent enabling it—are the various coastal engineering efforts by state and federal agencies, in the form of levees, canals, dikes, and groins, to stabilize coastal land for the purpose of providing firm, insurable ground for real property improvement.

For Native peoples, the island served as summer gathering place for celebrations and to harvest the whales that washed up on the beach or became stranded on sand bars. Marian Fisher Ales relates that when a whale came ashore, a powwow was held and the fins and tails were offered as sacrifice.[3] The meat was considered a delicacy. From contact and settlement through the resort era, rights to the beach related to the harvesting of resources. Long Island farmers used the Great South Beach for grazing cattle and harvesting salt meadow grasses; other uses included shipwreck salvaging, slave running, and whaling.[4]

Resort development on Fire Island grew rapidly after the Civil War with the construction of hotels along the beach. Subdividing land for streets and house lots began after 1900. By the 1960s, there were over eighteen communities and private resorts on the island. But for lack of easy access, their growth could have continued until the entire island was built up.

Robert Moses, the New York highway building mogul, had a solution to the access problem. Moses built an Ocean Parkway as part of his Jones Beach project in 1929–30 as far east as Captree Island. Moses proposed an extension along Fire Island all the way to Southampton, following the September 1938 New England hurricane (still unsurpassed in its destructive impact on Long Island, Connecticut, Rhode Island, and Massachusetts).[5] The parkway would stabilize the barrier island and its mainland protection but at the cost of thousands of summer homes in the Fire Island communities. Moses revised the parkway proposal in 1944 and again in 1962, after a northeaster that damaged the dunes and destroyed forty-two houses. Fire Island summer residents and their allies in the Suffolk County legislature blocked the project in 1938 and remained opposed in 1962, when their efforts were met with a favorable policy toward national seashores. Since the 1930s, the Department of the Interior had been looking to acquire land to protect the nation's coasts from private development. The communities welcomed a "national seashore" on Fire Island as a mechanism that could permanently thwart the Moses parkway. Fire Island National Seashore was established in 1964 in part to preserve the island's existing roadless character.[6]

Island Life

Fire Island accommodates dense settlement but without the usual complement of roads and cars. The island is loaded with public lands and a continuous beach uninterrupted by channels, jetties, fences, or other obstructions. The island has the exceptional nature-sanctity of an intact world of beach, ocean, and bay graced by the houses, gardens, and walks of a series of communities.

Fire Island has the inherent advantage of one of nature's most vivid landscape types: the land-sea edge comprising beach, ocean, and horizon, and, in back, the calm expanse of Great South Bay. This is a walking, being-in-the-world kind of place, delightful to visit on a summer day. Arriving at the dock after the ferry ride across the bay amid the bustle of greetings and the loading of baggage onto the pushcarts and cargo bikes lined up near the dock, visitors walk across the narrow island through cozy streets sized for the human body, then up the stile over the dune and onto the beach. Unlike parks and other places where people do not generally live or work, the Fire Island communities offer the unusual experience of a built-up residential landscape where streets are like sidewalks and everyone moves about on foot or on bicycles. Like other towns, single-family dwellings line up along the typically straight streets in the Fire Island communities. The beach vernacular street grid prevails here, as in so many other beach settlements developed incrementally and without any overarching plan, but with an important difference: Fire Island has no "Ocean Drive" or similar prominent lateral way alongside the beach. Long straight streets lead across the island from the ferry landings, interrupted only occasionally by cross streets, running parallel to the beach, but no wider or otherwise more important than the north-south trending streets. The predominantly north-to-south, cross-island street layout echoes the cross-bay ferry connections with the three ferry departure points on the south side of Long Island.

The Seashore serves visitors at its public areas accessible by ferry and private boat: mainly Sailors Haven and Watch Hill. The Seashore maintains the five-mile-long Fire Island Wilderness which has no facilities, and the Fire Island Lighthouse, which can be reached by foot after driving across the Robert Moses Causeway to the state park at the island's western end. It may be the case that the communities receive more visitation than the Seashore areas as some offer a typical summer beach community scene as well as the beach itself.

All the communities but Point O'Woods can be reached by public ferries. Cherry Grove is a fabled gay resort with charming elevated plank walks and a retail scene. Ocean Beach is the central attraction in the western half of the island, where visitors find shops, restaurants, and a guarded beach. Ocean Bay Park has Flynn's, a rowdy bayside establishment easily reached by ferry or private boat, located right alongside the locked gate of exclusive Point O'Woods. A day tripper cannot take a ferry to Point O'Woods or walk in from Ocean Bay Park, but one can visit by walking over the stile from the beach or from the eastern side via Sailors Haven and the Sunken Forest. Many of the Fire Island communities experienced a mid-twentieth-century period when "groupers" were predominant in the summer scene—that is, houses rented to groups of unrelated young adults prone to loud partying. Most, including Ocean Beach, Seaview, Fair Harbor, and Kismet, regarded themselves as having transitioned to "family communities" by the time of our research twenty years ago, but Ocean Bay Park still had numerous grouper houses. As one Seaview resident put it, "No family, once they fix up a gorgeous house, is going to rent it to twenty-five crazy people who are going to trash the house! Why would they do that? People love it here, they're house proud. Houses are now worth a million dollars here." There are, of course, owners who enjoy the income from group rentals as much or more than taking up residence themselves.

Day-trippers and groupers can be nuisances for the communities. The Seaview resident described:

> The scene among the *night* trippers—people who come over to Fire Island for the evening and they're drinking at the bars and they walk through the streets in hopes of hooking up with somebody and they're looking for a place to do it. We've talked about coordinating efforts to control all that activity from the late ferries on weekends and hired off-duty police, paying them with our hard-earned money to control all that activity, move people through between Ocean Beach and Flynn's in Ocean Bay Park. The night trippers want a good time with no rules. We're between Ocean Bay Park and Ocean Beach, where all the bars are. So every Friday and Saturday night, there's a constant stream of people coming through. . . . You know, if they don't get lucky or don't like the drinks down there, they come over to see what's going on here.

This resident went on to illuminate the conflict between family residents and the business community, who "get an incredible amount of business from the day-trippers. And the store owners in Ocean Beach love it." What she describes are the ordinary conflicts between sedate residents and party people that occur nearly everywhere. The conflict distinctive to Fire Island is the one between residents' interests and the Park Service's mission, baked into the legislation that established the National Seashore with eighteen community inholdings.

National Park Service Responsibilities and Concerns

The Park Service refers to the National Seashore as a "resource" or set of resources, a policy abstraction at odds with the homeowner and community residents' interest in preserving their property and its social setting in place. The mission at the seashore, as at national parks everywhere, is to set aside the natural system and its physical features and wild inhabitants from the real estate market, protect the delimited space

from development, make it available to visitors, and to interpret its significance. A fifty-year-old newspaper story illustrates the Park Service
perspective on Fire Island as a resource: "The realization that the island
is shifting rather than disappearing has led Federal officials to see erosion
as a less than critical problem, particularly since any program to restrict
nature would be inconsistent with the conservationist principles behind
the National Seashore designation." As the article explains, winter storms
skim sand off the beach and transport it to locales farther west. The state
park at the western end of Fire Island lengthens about three hundred
feet every year with sand deposited by this process of littoral drift. The
accretion left the Fire Island Lighthouse, built at the western end in 1826,
five miles east of the western tip by 1973. "Erosion may be harmful in the
short run," said James W. Godbolt, superintendent of Fire Island National
Seashore at the time, "but it all evens out in the end. What is taken from
one area is given to another—which, of course, may not be of much com-
fort to an owner who is going to lose his house."[7]

Fifty years later, the Fire Island Lighthouse Preservation Society reports the lighthouse as six miles eastward of the tip, the island having
lengthened by a mile since 1973. The tendency of storms to erode the
beach and shear off the foredune has taken many houses over the years
and eliminated the larger enterprises that once stood on the oceanfront,
including the Inn at Point O'Woods and an Ocean Bay Park restaurant
and deck popular in the 1960s. I don't know whether the lack of a lateral
street along the oceanfront is a coincidence or if the people who laid out
the streets in the Fire Island communities knew about the littoral drift.
Either way, the thinning of the island has shortened some streets and undermined front row houses, but unlike Venice, Atlantic City, and so many
other strands, Fire Island has no road or boardwalk running alongside
the waterfront to protect against or be undermined by winter storms.

The lack of a defining lateral public way facing the beach makes it
more difficult for the Park Service to identify and acquire the property
of houses destroyed by storms. In the rearranged landscape, people get
building permits for new houses. In the view of a Park Service urban
planner, new houses get their local approvals because, she says, the
towns of Brookhaven and Islip (massive jurisdictions the size of counties
in most states) don't really care about reviewing Park Service comments
on a building application. The planner also noted the tendency for small

houses to be replaced with large ones: "What used to be beach houses now become beach palaces—gigantic, ostentatious houses with swimming pools. Every time a little house goes down a bigger house goes up. And that causes people to hang on much, much more tenaciously to their investments. Beach houses used to be little disposable things. They've become these incredible, huge—they hang on very tenaciously, they have this God-given right called property rights, and until someone drags their ass into court . . ." For the communities, "protection" means keeping the ground they stand on intact, and for that to happen, the beaches and dunes must be made stable. Some communities have assessed themselves to "nourish" their stretches of beach, depositing sand to build up the profile of the beach and to strengthen the foredune. Some have built groins to slow down littoral drift. The federal agencies, for their part, take the long view of the natural ecosystem as whole. The communities are part of that ecosystem, but if individual houses come and go with tide and time, their existence, to the Park Service, is like that of shells on the beach—always there but perhaps not the same exact ones as last year. At the Ocean Beach historical society, a year-round resident, overhearing a conversation between this researcher and an historian, voiced a common grievance: "The National Park Service has made it very clear they don't want residents here at all. This is a national park. I've heard that for years and years, that they really don't want any of us living here." Sentiments like this magnify the complexities of federal oversight over the affairs of microcommunities of privilege into an all-out, blanket position: "They don't want us here."

Homeowners are a protected class in the United States with zoning to maintain the stability of their single-family neighborhoods from invasion by gas stations, townhouse apartments and other dangers, and by the mortgage interest deduction, federal mortgage insurance, and other rewards of homeownership. While all these advantages obtain on Fire Island, the authority of the Park Service to regulate driving, access, and other resident prerogatives can remind the adult residents of their younger selves, beholden to a parent's *because I said so* authoritative style.

In the same conversation, another resident related a more fully formed conspiracy theory: "Some say the NPS [National Park Service] wants to chase year-rounders off the island by making it so difficult for them. Then, once they're gone, there will be a calamity—a fire that takes out most of the summer houses. Then the park can acquire the property and make it all a park." Offline, Park Service staff can be equally hyperbolic about the communities: "Fire Island is a contradiction in terms of what the definition of a national park should be," said one. "Almost like a private gated community and the Park Service protects these gated communities, private enclaves."

Ecology and Wildlife Concerns and Protection

Others find more balance. The Park Service has a reputation for centralized services—interpretive schemes dreamed up at the Denver service center by experts with no grounding in the locality—and for transferring personnel around the park system. This may be true, but all the people we spoke with, other than the superintendent, were Long Islanders with considerable knowledge of Fire Island, its communities, politics, tensions, and issues. "Andy," the manager at the National Seashore's William Floyd Estate unit, on the Long Island mainland, expressed a personal devotion to the naturalist values of the Floyd family who donated the estate to the National Park System in the 1970s. Andy described the Floyd heirs as "very adamant that the place not be consumptively developed." As Andy explained it, the family once had a large north shore estate, in Smithtown, Long Island, which the county acquired through eminent domain to be a public park. "Yeah, the county took it and turned it into a mass consumption park. You can go there now and there's camping trailers and—you know, the integrity was removed from the property and it's just a mass usage park. That," he said, "was a bitter pill [for the family] to swallow." By his account, the Park Service's proposal to maintain the landscape intact, "maintaining the fields, allowing the woodlots to mature—no playgrounds, ballfields, that type of activity," resonated with the Floyd heirs. "Let's face it, they could have made a lot of money by flipping this place into private development, but their ethics wouldn't allow them."

The Park Service has preserved the Floyd estate intact, but without the connection to living family members or much of any other use, its hundreds of tick-infested acres on Moriches Bay stand full of wildlife but largely empty of human use. Nevertheless, Andy described how he had taken it upon himself over twenty-five years to tag individual eastern box turtles on the estate grounds, following, as he put it, the practice of one of the Floyds. Andy also managed the hay mowing regime so as not to disturb the turtles, an upland species threatened on Long Island by suburban development. This was not mandated, but Andy had such respect for the donor family's ecological values as to have adopted the family's practices. He claimed to have tagged 715 turtles.

For Park Service staff, who work for an institution whose purpose is to protect the natural communities of the bay and barrier island ecosystem, protection relates more to the National Seashore's mission than to the summer residents' goal of protecting their communities and private property from natural disaster and official interference. Some park staff had built careers that allowed them to live on the island the year around. "Jim," who in retirement volunteers at Fire Island Lighthouse, emphasized the island environment's affordance of living simply and in harmony with the natural cycles.

> When you live here year-round you become part of it, as far as working with the environment, dealing with the environment, or playing with the environment. When the fish are running you are fishing, you know, when the blueberries are ripe you are picking blueberries. When it is raining outside you may be reading or something. You become very much a part of that and your feeling of oneness, I guess, as a family you become used to that. You don't run down to the corner 7-Eleven, you don't do those things that normally are done. But you live and you change and become one, I guess is the closest thing I can think of, with the way you work. You know you have to go off to the mainland but you don't do it as often because, you know, it's a bit of a journey and it does take time from you doing what you want to do. Even though it becomes all part of it as far as going back and forth across the bay, getting to know the weather, you get to know when to go, and everything becomes part of a balance too. That's what is so great about it.

Jim related a kind of pastoral of older ways living close to the land, attuned to the next fishing or hunting possibility.

> Water Island still has the flavor of what it used to be. Oakleyville is another area, between Point O'Woods and Sailors Haven. It brings back fond memories of when we first came to Fire Island; that's where I lived

year-round. The people there were very much the same way, part of the season, part of the landscape, part of the area. They had history there, they knew the history, they knew what plants were here. They had a familiarity with it, they had a oneness with it. They knew when the scallops were there, they knew where the clams were. That's what they ate. This is the beach plum shrub that you have with a little vodka, you know.

This is in contrast to the homeowner in an overly large new house built over the dune who demands government action to protect his asset and privilege. Managing the resource involves setting limits on the use of personal motor vehicles, access to areas where protected species are known to be nesting, acquiring house lots to prevent development in ecologically fragile zones, and similar restrictions. The National Park Service cannot keep people or residents from becoming more removed from the natural cycles, but it can and does work to protect the natural systems and threatened and endangered species from being despoiled by the tendencies of people operating within an exploitative political economy to prioritize self-interest at the expense of the common good.

Lived Experience versus Official Expertise

An architect who lived on the island beyond the usual summer interval criticized the lack of horticultural improvement in the park's undeveloped spaces. He thought the communities had become so much more beautiful over the years for all the gardening efforts of residents. The undeveloped spaces in between, by contrast, were bleak for lack of effort. This limited view reflects an important aspect of engagement with nature and natural systems: just as fishing produces knowledge of fisheries and the systems that support fish, through gardening, planting, tending, and caring for plants and soils, the gardener gains a certain knowledge that others acquire through professional training. The architect may not engage in gardening, but he appreciates his neighbors' gardening efforts and has perhaps a closer range experience of neighboring front yard gardens, as he necessarily walks through the narrow lanes of his community within sight and scent of everyone's horticultural efforts—unlike the suburbanite moving in a climate-controlled vehicle past pallid lawns and diffuse greenery. The resident also talks with other residents, as people do, telling and hearing stories of life on the island, and some of the stories touch on gardens and plants and beautification efforts and all the other things people talk about that produce an understanding of, a *position* on, the community and its landscape and setting. His scorn of the Park Service's lack of horticultural improvement in the natural areas may strike an environmentalist as uninformed, but it stands firmly within the long-standing cultural view of nature as a material to domesticate in making settings for productive social life, rather than to leave alone to be admired for its own sake.

For the architect, the walkways and beaches are foremost a social space that reinforces community belonging: "This place is a completely outdoor environment. This environment lends itself for you to be out, and when you're out everybody sees you. It's kind of a strange place that's isolated but has a social environment where everybody knows everybody else. Maybe not well, but you do know them. On the other side, you might know twenty people and that'll be it. I know a thousand people. It's an outdoor environment that allows people to be who they are outdoors, outside." Like others on the island, the architect felt he understood the ecological system better than the federal experts in the Park Service and the Army Engineers, who, he said, had been talking about and studying the problems of littoral drift for forty years but not taking action. Even without all the stalling, he thought the federal government would never pay for an undertaking of the scope residents think is needed. "We all know the problem. The inlets are the problem since they opened in the '60s. That's our sand; That sand belongs to us!" He went on,

> They should have dumped old subway cars to stop the surge from coming onto the beach. Environmentalists are not happy, they say we're destroying habitat on the dune. My answer: If there's no dune, there's no habitat. The Seashore on its property does nothing. The dunes will never reform unless you help them. "Leave it au naturel—leave it the way it is," they say. They feel we should do same, let the houses go. They say it's a playground for rich. I don't see lot of rich people out here. Mostly middle class; the day-trippers certainly are not rich. Their posture: We're evil people. Unfortunately they're fucked because they can't do anything. Instead of working with us—because this is a phenomenal place—they fight with us.

While the architect and many other seasonal residents would protect the beach to protect their real estate, protection for the Park Service involves more activist management of seashore resources than in earlier periods of the national seashore. A civic leader in one of the western Fire Island communities observed that the Park Service had been "asserting its stewardship, which it didn't use to. Now there's much more environmental pressure off the island to maintain the fragile beach, especially the piping plover and the sea amaranth. Also litigation by off-island groups who object to the Park Service allowing this, not restricting that, letting cars on, not letting cars on. And this is nationwide that the Park Service is the focus of a lot of litigation—primarily by people who are vociferously pro-environment." This civic leader strove to offset the perception of community residents as a self-interested elite at odds with the National Seashore's mission. Rather, "the thing I keep pushing is that summer people are stewards of the island just as much as the Park Service people." Why? Because the summer residents have the experience of living on the island for days or weeks on end, sometimes in different seasons, some with family histories going back to the beginning of residential settlement. For those with longer histories, the island isn't an isolated place you reach after a drive or a train trip through featureless space but part of a larger sociocultural environment that knits mainland to island. A Babylon resident and onetime neighbor of Robert Moses, himself a Babylon resident, told a story of getting out to Fire Island before the ferries.

> Even now it's hard to get to Point O'Woods in bad weather, but then it was mostly by sailboat—later on a steamer, but poor ferry service. And no plumbing, electricity—And it was rustic living. Even when my grandmothers—both widows—went to Point O'Woods—one from here [Babylon]—they had all these baymen who did fishing and stuff and they would just get one of these big sailboats and the man would sail them across. My grandmother, three children, the sewing machine, mosquito netting, supplies of all sorts—kerosene—that was how they got there. Then some wagon to get them up to the house. Like pioneers—no doctor unless a resident happened to be one. My grandmother went when she was six years old—she was born in '88 and the house was built in '94.

This resident described relationships with the flora and wildlife people gain as residents:

> I could give you the names, practically—about eight or ten regular fishermen, or more even, that regularly catch bluefish particularly, and maybe stripers some, but I know they catch bluefish. Then they also fish off the dock—kids do, particularly, because kids are not so good at surf casting. They get flounder and weakfish and blowfish. And then there was clamming. We always clammed. Went in the bay, not off the dock where it's deep, but further down. There's a walk that goes down to where you can just step into the bay and walk out. And you tread for clams—you feel them, of course, and the clams lie sitting up this way, so what you feel is the sharp shell. And, of course, you feel a million other things that you think could be a clam, broken glass, something like that, and then you go down and just pick it up. My husband had a clam rake—not the kind you use on a boat, but just this kind, and he would just do this. It's a long pole with a metal cage sort of thing, and just do this about four times, then pull it up and see what you have. Take out the clams, if any. . . . And we always did that, my husband and I, and all the kids did too, and sell them. And everybody knew how to make clam chowder. And when they got older the kids would just go around door to door and sell them. But now the clams seem to have just disappeared. They're very, very scarce. You rarely hear of anybody getting any clams.

In a similar vein, an older year-round resident of Ocean Bay Park recalled an earlier era free of regulatory oversight: "Before the bridges were in, when nobody was paying any attention to us here. You know—like [Native Americans], we could roam. It was a great feeling. We would go up to the oceanfront and would build a fire. We were smart—we dug holes into the ground, and then we would build our fires. We took the water from the ocean and put it onto the stove. We cooked corn, and we made chef's grills, made hamburgers and hotdogs, or steak, or whatever we wanted. Again, there was nobody to bother." Such similes—likening themselves to Native Americans and pioneers—ground these community narratives in the tradition of hardy risk-takers who, through their efforts, built community on and in harmony with this unusual island environment.

These narratives point to an accentuated self-identification with the island that comes with extended residence. With such place identity comes knowledge of its systems, possessiveness, romanticization of past ways of life, determination to preserve what remains of it, and resistance to the federal authority that prioritizes the ecosystem and regards with wary suspicion the resident's defensive attachment to island residence. The civic leader said that preserving the island communities

> should be in the General Management Plan, part of the mission of the Seashore—that on the edge of a huge metroplex is this—"enclave" is the wrong word because it has elitist overtones—this *pocket* of small-town life that's been preserved because it's on an island, a sandbar, in the Atlantic Ocean. It doesn't have vehicles, access is not easy, and so we're kind of a backwater. If you want to make a metaphor, it's kind of like an Amazon tribe in the jungle, far removed, and so it's maintained a lot of its traditional way of life. And we're fiercely protective of that.

Conclusion

Sixty years ago, the Fire Island communities faced the prospect of an Ocean Parkway that would preempt oceanfront real estate and make the island more widely accessible. They welcomed the National Seashore to eliminate the parkway threat. The transaction favored the communities, who got to keep themselves whole, retain fee simple possession of the underlying land, and maintain the island's roadless character. A National Seashore with communities bought out and removed and the natural landscape of dunes, coastal forest, and salt marsh restored, with provision of facilities for recreation, environmental education, and so on, would have restored Fire Island to its former status as a commons, serving the broadest possible bundle of public interests. The compromise that created the Seashore but preserved the physical character and underlying individual fee simple possession of real estate within the communities left the Seashore in conflict with private interests that limits the Park Service's ability to provide a truly public environment and to protect the land-altering dynamics of erosion and accretion in a barrier island environment.

One can imagine a solution more favorable to the public interest that could still have won the necessary political support. The authorizing legislation could have adjusted the bundle of rights held by the various stakeholders in the limited and fragile resource system that constitutes Fire Island, perhaps providing for acquisition of the underlying land, with provisions for residents to remain in place with ground leases, possibly to sell or otherwise transfer the improvements to others. The Park Service would be owner and manager with authority to restrict development, limit the upsizing of houses, restrict the horticultural prerogatives of residents as concerns over invasive species grew, remove houses and streets as necessary to reconfigure the shoreline or interior landforms to better manage the resource to absorb tidal flooding as sea levels rise, and so on. Allowing residents to remain in place after 1964 on ground leases could have attenuated the residents' sense of deserving what few others have while allowing the cultural lifeways on the island, in the bay, and on the south shore of Long Island to persist.

The voices heard in this chapter reveal conflicting versions of protection but also commonalities that point toward cooperation. Residents, homeowners, and Park Service officials all acknowledge Fire Island as a special place that comprises a set of natural and social resources to be protected. No one is asking to build roads and armored seawalls. Most communities join with the Park Service in opening the island's streets and beaches to day-trippers. All acknowledge the desire to protect the barrier island environment of beach, dune and bay bottom. All value the opportunities for fishing and everyone wants the bay to be productive of clams and oysters once again. The residents' knowledge of the environment gained through experience and tradition can enter into a new dialogue with the water resources and wildlife protection experts in the Park Service. The experts can listen to the residents and come to appreciate the place of historic seasonal community lifeways in the island's natural systems. From the experts, residents can come to appreciate the scalar-temporal mismatch in some of their arguments—for example, offering the nature-culture balance or the hardy settler myth of olden times as arguments against the island-wide management initiatives driven by present-day challenges. The Park Service has to protect the seashore from the effects of a much larger human population in the region, lots of day-trippers and boaters, and from a securitized real estate economy unknown in the 1960s that transforms quaint prewar beach houses into substantial structures that make far greater demands on the island's sewerage, stormwater management, drinking water, and energy resources. The demands of some residents for one-off interventions in the natural systems, like piling up subway cars in the inlet, building a groin across the intertidal zone, or dumping dredge spoils to build up the beach, are out of scale with the dynamics of change at the land-sea edge as waters warm and sea levels rise. Seeing the protection of natural plant and soil communities in the common areas of the Seashore as an abdication of horticultural responsibility disregards the devastating impact park and garden improvements can have on indigenous plant and wildlife communities.

It is no wonder that residents are fiercely protective of their summertime way of life. The privilege of owning permanent twenty-four-hour access to a rare coastal environment only fifty miles from the city of New York does not escape the residents' notice. Haunted by the specter of a calamity that would at last give the authorities the opportunity to kick them all out, the residents spin defensive narratives in which it is they who know best for the resource, they who should be partners with the Park Service as managers of the resource, and they who deserve to sustain their atavistic small-town paradise. The interpretive mission of the Seashore can engage these impulses of stewardship while touching on historic lifeways, the clammers and oyster harvesting baymen, the Indigenous powwows and use of distinctive Great South Bay clams for wampum, the thousands of shipwrecks and related scavenging, the origins of the Coast Guard in the Life Saving Service, the understanding of the Great South Bay and Beach as a commons continuing long after European settlement, the transition to seasonal occupancy, the LGBTQ+ history at Cherry Grove, the Chautauqua origins of Point O' Woods, and the diverse environmental knowledges residents acquire through experience and lore. None of these cultural traditions relies on fee simple possession of island real estate. As environmental pressures grow, the property rights structure that unduly favors the homeowners' prerogatives can be reexamined and potentially reconfigured to serve the common interests of all stakeholders.

NOTES

1 This chapter is based on data gathered in a study by the authors in 2003–2005 under a cooperative agreement between the Public Space Research Group at City University of New York and the National Park Service. Setha Low and Dana Taplin, *Final Report: Ethnographic Overview and Assessment, Fire Island National Seashore* (New York: Public Space Research Group of the Center for Human Envi-

ronments Graduate Center, City University of New York 2006), http://npshistory.com/publications/fiis/eoa.pdf.

This ethnographic overview and assessment study was intended to identify cultural ties between places and people that have endured over time and transcend recreational phenomena. Ceremonial grounds, sacred sites, and other such places once identified are considered ethnographic resources that should be protected. The Park Service commissioned the study to inform the development of a new General Management Plan for Fire Island National Seashore.

The methodology involved documentary research in local libraries, museum collections, and online sources. Setha Low, Dana Taplin, Andy Kirby, and Mara Heppen of the Public Space Research Group conducted key informant interviews with residents, Park Service people, historians, and others. Our Public Space Research Group at CUNY had pioneered Rapid Ethnographic Assessment Procedures (REAP) methodology at other sites and infused some REAP methods into the research, including participating in community events and random interactions with residents.

2 Andrew W. Kahrl, *The Land Was Ours: How Black Beaches Became White Wealth in the Coastal South* (Chapel Hill: University of North Carolina, 2016).

3 M. Ales, "A History of the Indians of Montauk, Long Island," in *History and Archaeology of the Montauk*, 2nd ed., ed. Gaynell Stone Levine (Riverhead, NY: Suffolk County Archaeological Association, 1993).

4 Laraine Fletcher and Ellen R. Kintz. *Historic Resource Study, Fire Island National Seashore, Long Island, New York* (n.p.: US Department of the Interior, National Park Service, 1979). Fletcher and Kintz offer a thoroughly researched general history of Fire Island performed under contract to the National Park Service that covers whaling, maritime industries, fishing, agricultural uses of the beach, baybottom oystering, and the development of summer communities on the island.

5 "The Great New England Hurricane of 1938 was one of the most destructive and powerful storms ever to strike Southern New England. This system developed in the far eastern Atlantic, near the Cape Verde Islands on September 4. It made a twelve-day journey across the Atlantic and up the Eastern Seaboard before crashing ashore on September 21 at Suffolk County, Long Island, then into Milford, Connecticut." National Oceanic and Atmospheric Administration and National Weather Service, "NWS Boston - The Great Hurricane of 1938," www.weather.gov.

6 Christopher Verga, *Saving Fire Island from Robert Moses: The Fight for a National Seashore* (Charleston, SC: History Press, 2019).

7 Robert E. Tomasson, "Ocean's Pummeling Reshapes Fire Island," *New York Times*, July 7, 1973, 261.

8

East River Park, Resiliency Politics,
and the War on the Trees

BENJAMIN HEIM SHEPARD

Public space is always shifting in New York; trees come down, condos
and developer plans rise. No place is this dynamic starker than in East
River Park, where urban policy involves cutting down trees to save the
city from flooding. This strange outcome is a symptom of a larger policy
conflict between a developer-friendly approach and a "consensus plan"
built around the strengths of the urban ecology, favored by the community. In the fight for East River Park, battle lines were drawn between
anarchist gardeners and technocrats, poets and planners, for old trees
versus new turf. There were users who liked the park just as it had been
since the 1930; there were planners who imagined something newer and
shinier, with less green grass to manage; performance artists clashed
with construction crews; tree huggers locked themselves to block jig-
saws; mayors awarded lucrative construction contracts to companies
to tear things down; people in favor of abundant uses of urban space
clashed with supporters of neoliberal urbanism; preservationists fought
against architects who were ever drafting plans to tear down the old, hire
contractors, and build the new. What was not there were policymakers
able to bridge the gap between opposing forces and interests.

After Hurricane Sandy, October 22 to November 2, 2012, the city
opened a competition led by Rebuild by Design to redesign the East
River Park to withstand rising sea levels. Lower East Side community gardeners, public space supporters, artists, professional designers, academic researchers, environmental scientists, some twenty-six
community groups, and teams of architects came together to initiate a
dialogue. The result of this community engagement was a "consensus
plan" that preserved the park and prepared for climate change. This
consensus plan proposed a planted berm along the park's western bor-

der at FDR Drive as a flood wall; it left most of the fifty-seven-acre, one-mile-long park—with close to a thousand mature trees and a wetland absorbing water—intact.[1]

Late in the process, the city rejected that plan in favor of a seawall that would pave the park, thus removing the trees and the community space, elevating the whole park, and covering it with ballfields. The East Side Coastal Resiliency Plan (ESCR) "used a series of berms, flood walls, flood gates and raised parklands to create a continuous 2.4-mile barrier to protect 110,000 residents of the Lower East Side in Manhattan from future coastal and tidal flooding."[2] ESCR was twice as expensive as the consensus plan and favored more concrete and synthetic turf while destroying the current park.[3] (Similar turf and microplastics installed in remediated areas as part of flood mitigation and brownfield cleanups, such as in Red Hook, Brooklyn, are ending up in storm drains and then waterways.[4])

All this was to take place in an area where people came together and found community, particularly during the pandemic. The following traces the transformation of this contested space and asks "What happened?" This is a story of resistance—of rallies and speak-outs to save the park and of testimonies by defenders as well as opponents of the plan for the park. This chapter considers the ambiguous framing of the organizers calling to save the park for the users, protect the community from the sea, and to leave the trees. It contrasts the different solutions proposed to protect the park from rising seas and how they developed and evolved.

This chapter is also about the park users and how they view the park in the context of the neighborhood, the stories they tell about it and the trees, their organizing to fight the new plan, and their failure to win support for their efforts.[5] As the gap between politicians and community grew, the story of East River Park came to be a part of an ever-expanding narrative of cleavages and clashes between planners and environmentalists—one of the countless battlegrounds over the privatization of public space,[6] over commons ever under threat.[7] To situate the story, I describe the setting and background of the conflict and the methods I used to investigate the issues, offer my findings (interviews, transcripts, meetings), provide discussion and analysis of the problem, and come to a conclusion about what can be learned from this case.

This is not a beach but a park on the water; it is an urban space contending with flooding. Geophysically, this chapter is different from the others in the section. And yet, the trees are key to keeping the land edge stable here. They do a better job than asphalt and a wall because their roots are alive and deep and they can grow with the shifts in currents and water levels that come with climate change. The connection between plants and water remains dynamic. The case of East River Park illustrates the kinds of politics that occur at the community and city level and why it is never simple to "save" a coastline that is in danger from climate changing and water rising.

Setting, Background, and Methods

Through street ethnography, this chapter investigates urban activism, expanding on narratives about the contested nature of public spaces. This reflexive case study builds on multiple data sources including my voice as an observing participant in activist groups supporting green infrastructure, discussions with other participants, and historical accounts.[8] Through participant observation, this chapter makes use of the researcher's experiences, thoughts, and reflections as subjects of consideration in and of themselves.[9] This form of research is useful for considering the uneven, ever-shifting politics of climate science and the complex notions of knowledge production, urban behavior, and political participation, as well as the messy process of translating knowledge into action as the groups covered here have done.[10] At its essence, this is activist ethnography. Reflecting on the very debates about what we know, how we know, and how we act upon information, such engaged ethnography has the advantage of enacting reflexivity. It highlights a gap between those fighting to save the trees at the water's edge and those, secure in their power, fighting to save the NYC traffic routes. For the former group, this piece of land is a place; for the latter, a piece of a transit and real estate puzzle. As a postmodern ethnography, this chapter does not shy away from the often contradictory, intricate, and even entangled nature of such research; rather, it embraces a creative plurality of voices and points of view.[11] In essence, it offers a description of the setting, how the park is (or was) used, the groups and people who used the park as it was, and why park users are so invested in a community plan to protect the park in the face of flooding.

East River Park has long had a place in the public imagination. It opened July 27, 1939, reshaping a former shipyard into an area for ideas and leisure. An outdoor amphitheater followed in 1941, open to the public for performance, just south of Grand Street. This use was against the wishes of Robert Moses, who opposed free public theater. Over the years, Joe Papp's Public Theater staged free performances of Shakespeare there. Five decades later, the members of the punk band Nirvana posed in front of the amphitheater. Around that time, a compost yard took shape and this "urban ore" made the park and its trees more lush and beautiful each year.

For many, this was a free space to jog, watch birds, get away, breathe some fresh air, meet a friend, join a picnic or dance party, or go on a naked bike ride. East Village poet Jim Carroll's diary highlights the sentiment: "Fall, 1963, we played tonight. . . . It's Friday night and all we wanted to do was go to the East River Park and get drunk, do reefer and sniff glue. And that's exactly what we did."[12] For Carroll and company, this space represented something distinct and liberatory, a place of wonder and discovery, a place to be oneself.

It certainly was for poet Eileen Myles. "I don't know when I first went down there," wrote Myles. "I moved to the East Village in 1977. . . . We would walk our Millers down to the park . . . and the night would begin . . . made my sanity great . . . this park is for everybody."[13] Of course, part of the appeal of the park and its sense of place was the open access to trees offering shade and solace. Trees lined the park, many dating back to those early days of the Works Progress Administration, swaying with the breeze by the river.

Trees and Urban Space

Hurricane Sandy left parts of Lower Manhattan underwater. The surge hit October 22, 2012, from 8 to 10 p.m. Water breached the West Side Highway, with waves as high as fourteen feet.[14] Trees went down throughout the city. Water in East River Park receded within a day or two. Trees outside the Lower East Side Ecology Center were still suffering after branches had been ripped from them during the storm. Many were toppled, leaving jagged open skies where branches used to breathe. The top half of a willow had fallen inside El Jardin Paraiso, a community garden on 6th Street and Avenue D. "Save the trees," chanted those on procession through the neighborhood after the storm. "Who will speak for the trees?" wondered a protester, carrying a sign referring to the Lorax. The city started cutting down trees.

"Deep ecology reminds us we are connected to everything," J. K., a garden activist, told me. Gardens and trees are part of something larger, J. K. went on, linking their roots with a web of life as a living, breathing force. Absorbing flood waters, cleansing the air, removing carbon, releasing oxygen, trees are critically important in coping with climate change; as breathing ecosystems, they support biodiversity, mental health, and urban ecology.[15]

Still, the city scrapped the "consensus plan" that saved the trees, in favor of ESCR, which was more favorable to motorists, taking down one thousand trees to rebuild East River Park.[16]

East River Action: Observations and Findings

Understanding community activism and its ups and downs is hard to capture except through engagement with the many varied, competing, often intersecting narratives. In this section, I try to give a sense of the hard work and frustration that went into first creating a consensus plan within the community and then fighting with the city to assert our community vision and rights.

Presenting this struggle in detail gives a better sense of the politics of fighting to save a coastal site (and especially the trees) that was loved and enjoyed by the community. Our sense of loss is part of this politics as well as part of the ongoing struggle with the city to put people's preferences and everyday use first rather than technological solutions to ecological problems.

On Sunday, April 18, 2021, I joined park supporters at Tompkins Square Park for a march to East River Park. We gathered in a semicircle around Hare Krishna, an elm tree dating back to 1873, next to the spot that the Bendy Tree, which had been bulldozed by the city after Sandy, had once lived. Advocates from past garden battles—poets, zoning opponents, and environmentalists—were all there. Eileen Myles, Emily

Johnson, and Harriet Hirshorn of East River Park Action framed the march as an "urgent exercise of free speech and outlet for the immense frustration we feel at the betrayal by public servants, i.e., our mayor, our city council member . . . and all who are engaged in supporting the farcical, greedy and short-sighted version of flood control known as ESCR." Myles and company explained,

> We had worked with city agencies since 2012 to come up with a strong green approach to coastal resiliency. . . . We supported the original . . . plan with its rolling berms that did not require the destruction of an 82-year-old park with 1,000 trees . . . a paradise of growing things and birds and fishing areas. The park is a multi-racial, multi-class, multi-age dream. It is a living symbol of the New York we love and treasure. . . . The glamorous and synthetic plan the city promises will in essence be a concrete wall along the river. . . . If the city says it will take five years, it will take ten. . . . There will be NO interim flood control in that time. If a storm occurs, then it will be a muddy mess. . . . We demand a stop to the East Side Coastal Resiliency Plan now. . . . The tall shiny buildings all over Manhattan are empty, this park is full, the evidence against this plan is life itself. And we march as life itself today to demand that we prevail and stop the plan.

March and rally, write letters and call representatives the activists did. Thanks to their continued engagement, park defenders did win the phased construction of the ESCR.

Opponents argued ESCR was an example of shock doctrine used to justify privatization and neoliberal urbanism after Hurricane Katrina in New Orleans. The same thing had been happening in New York for a generation.[17] The process takes place through the manipulation of crises to forward unpopular, previously unacceptable policy proposals while impacted communities are focused on the problem at hand, creating a mess that only private developers can fix through development.[18] In this case, the poststorm environment opened opportunities for struggles over what climate change adaptation should and could look like in urban waterscapes.[19]

East River Park was only one of several frontlines in the war on green space in New York. In fall 2021, I joined opponents of the Governor's

Island Rezoning at a press conference opposing that plan, which would open the pastoral half of the island to high-rise development. "What will the space become?" said Roger Manning. "Will it be better off?" Recall Penn Station, an architectural gem that was destroyed in 1963 in the name of progress, said several others on hand.

"We're not asking for more green space," said the next speaker, "just to keep the green space we have."

"Actually, I am asking for more green space," I said, beginning my testimony that connected the struggles for green space from East River Park to Elizabeth Street Garden to Governor's Island, all egalitarian spaces open to all under threat from the city.

"What are we going to do if the park is destroyed?" asked some kids in the park. "We come here every day."

By September 2021, the fight was accelerating daily. The city council had approved the ESCR, along with other rezonings. Mayor Bill de Blasio's term was ending. Advocate Sarah Wellington put out a call for park defenders to meet. "A thousand people for a thousand trees. Stop a tragedy. You can't take back a tragedy," said Wellington, calling for an oversight hearing. Over the next few weeks, park defenders would converge on city hall. "It's up to us to stop it," said Wellington. "To block the bulldozers, to call for state oversight. All the politicos are scoundrels. Let's get to know each other and strengthen our networks from within."[20]

Each day, I trekked to city hall, reading poems with other park defenders, making common cause with the high school kids and college professors, writers, and nurses congregated there, trying to get the word out.[21]

"My heart is with the forests," said J. K., while locking herself to a tree at City Hall Park in early September. Later, she reflected, "I was locked so close; I was almost with the tree." She said, "All day, we kept calling Corey Johnson's office, asking for an oversight hearing, asking what was going to happen to the trees."

Talking with passersby, activists reviewed reasons to support the "consensus plan" and its design. "The system plans to make more room for cars and cement, but underestimates the value of the natural resources, which cool the climate, hold precious water, and provide habitat for innumerable creatures, herbs, and flowers," said Erik McGregor, a local activist.

"The East River Park is a source for migration, for birds coming," said another activist. I forgot to write down their name. "If you kill the trees, you kill the nesting site for the birds," they continued. "We lose the esplanade. We lose the composting. We lose the trees, making the city hotter and dirtier." Cities need more trees not less, and more green infrastructure not less, contended activist after activist.

Sandy Charles stood up at one of the speak-outs: "Our relationships change when we become intimate with a space," said Charles. "Especially one we share with a community of people. Strangers become friends and an extended family. People take care of each other. . . . To feel the loss of a tree, let alone hundreds, the disease this creates, the disease the city planners and politicians bring to the table, not respecting mother nature, or the people that rely on this space for our health and growth, we say no to this dangerous plan."

In late September, Sarah Wellington and Eileen Meyers confronted the speaker of the city council, in a rare face-to-face encounter with a decision-maker. Passing the buck, Corey Johnson said it was not his district, and that he needed a formal request from council members Justin Brannan and James Gennaro. We called both councilmen, trying to get a formal request for a hearing.

Saturday, October 9, 2021, we showed up at every entrance of the park with hard hats and caution tape to stage a mock "shutdown" from 10th Street down to Corlears Hook. This fake closure was a way to illustrate the very real planned demolition ahead. "Park's closed . . . unless you do something about it," we tell those trying to get to the park. Some stay to talk. Others keep on walking.

In November 2021, park defenders, including myself, joined climate activists for the COP26 in Glasgow, Scotland, for the United Nations Climate Talks. Eco designer Wendy Brawer, of Green Map System, was there all month. At the Fridays for the Future march, Brawer told me about her time at the COP26, linking her climate activism with the seventeen principles of environmental justice calling for cities to be rebuilt in "balance with nature." Throughout the COP26, she told interviewers about East River Park and the need for the 30 by 30 principle, that 30 percent of all land and oceans be set aside for nature by 2030 (now in effect as of December 2022). World leaders at the COP26 emphasized ending funding for fossil fuels, cutting methane emissions, protecting

Indigenous rights, and fighting deforestation. Despite this, the City of New York was moving forward with plans to kill one thousand mature trees in East River Park, said Brawer. The gap between commitments and what was needed to avoid more than a 1.5°C rise in temperature and subsequent floods, droughts, storms, and tides was only becoming more and more vast.

At 7:15 a.m. on December 6, 2021, I met the park defenders at the amphitheater in East River Park. Construction trucks were everywhere. Fences were going up, the drilling was beginning, and the trees were next. Among the chaos of jackhammers and saws buzzing down the trees, objections to the plan took shape as gestures of civil disobedience, dramatizing people's various concerns, some ecological, others community-based, regarding the plan process or around social justice concerns. Lack of social justice was felt in both the procedure (everyone was asked to be involved but then ignored) and disappointment over the outcome. These themes are central to the politics of preserving public space for the people—not just coastlines and beaches.

"You don't destroy something to save it," said Kathryn Freed, former justice at the New York State Supreme Court, voicing a process concern. "How come no officials are out fighting for us?"

"More concrete is not the solution," screamed a young woman, voicing an ecological concern as construction crews brought in their jackhammers.

While standing and watching the cops line up with their plastic handcuffs, a teenager told me he came here every day during the pandemic—it was his safe place to hold it all together—voicing a space equity concern.

Councilmember-elect Chris Marte dropped by as the bulldozers were about to begin. "This park was a sanctuary when I was growing up," he said. "The whole world is watching us destroy this park because the city doesn't want to lose one lane of traffic. We can do better." It's like the revenge of Robert Moses.

"Whose Park? Our park," activists chanted, watching the destruction, while the violent sound of jackhammers filled the air.

"This was where our kids learned to ride," said Allie Ryan, who ran for city council in 2022 to try to save the park, visibly shaken, tears welling up in her eyes.

In their own ways, each of these observations highlights specific dynamics of transportation engineering and how it affects waterfront neighborhoods, specifically the adherence to anti-ecological, out-of-date, cement-reliant engineering to deal with water-land edge infrastructure. Bikes versus cars and urban space, the case of East River Park highlights the working of neighborhood park spaces as distinct forms of land-water edge spacial politics that can be softened or hardened, depending on the sociopolitical dynamics.

Police warned us that we would be arrested.

"You have to stop," said Virginia, attempting to block the machines, before she was arrested. She was one of many of those disrupting the construction arrested that day.

By the end of the day, activist lawyers had won a temporary restraining order, halting the bulldozers.

Later that night, a few of us met at the Lower East Side bar the Magician to unpack the debacle at East River Park and tell stories about jail time for protecting the trees.

"I was charged with trespassing and go to court in a few weeks," Virginia told us.

"Why would NYC forgo a model that emphasized resilience and biodiversity that cost less?" wondered Wendy Brawer.

Oh yeah, the mayor has debts, Eileen Myles reminded us. "When the last tree has been cut down . . . the last river poisoned, only then will we realize that one cannot eat money."[22] Pave a park? Put up a condo? What comes next? Of course, the planet is resilient; it can withstand a lot of human destructiveness. But there are limits. Climate grief grips many of us.

The next day, the destruction started again, the jackhammers pumping and the trees going down (despite the temporary restraining order). Each day, we met just below Houston Street and the park and watched the chainsaws take down tree after tree. And the conversation expanded.

"Save East River Park before it happens to a park near you," said one of the park defenders, standing in a circle with the other activists, the trees falling in the distance.

"Bury the highway, not the park!" activists screamed, fenced off from the amphitheater where Joe Papp's Public Theater had staged free performances, watching its demolition.

"How many trees did you kill?" Sarah Wellington yelled directly at the workers.

"Let's have an earth riot," gushed Bill Talen, a performance artist also known as Reverend Billy.

"All through high school we went to East River Park," said my daughter, back in town from college, watching.

"What do you think, Ben?" asked Talen.

"Plant a new tree, build a new green space," I replied, while bull-dozers were actually pulling the seats out of the beloved amphitheater. Throughout the efforts to beat back the bulldozers, storylines shifted from an emphasis on Hurricane Sandy as actor to groups of noisy chain-saw-wielding actors, each triggering another phase in various simultaneous conceptualizations of the crisis followed by yet another version of the story.

New ideas took shape in each meeting outside the construction zone as activists hatched plans. Many argued the city was in violation of alienation law. This is a precedent requiring state approval before parkland can be transferred for alternate uses. When part or all of parks are being transformed, activists look to alienation law to slow things. Arguing the ESCR transformed the park, East River Park activists sued the city.[23] State senators agreed with the activists on alienation—always off record. The courts ruled against the activists, offering a green light to tear down the trees.

"We lost our appeal and we lost," posted East River Park Action. "This decision sets a terrible precedent for all parkland. . . . All they have to do now is tack on some park-related excuse to whatever they're doing and it will not need to go through alienation or state oversight. They could put a building in a park and say it's for environmental research . . . and it will be ok."

"This is not a single-issue struggle because we do not live single-issue lives," said one of the speakers at a speak-out, paraphrasing Audre Lorde.

"The struggle is intersectional," said Ty, a New York high school student, who joined the rallies almost every day.

By mid-December 2021, the *New York Times* published a story that framed the conflict in terms of race and class.[24] The *Times* article pitted New York City Housing Authority (NYCHA) residents who live across from the park against East River Park activists in the same area, suggesting park defenders were out of touch—much like Rudy Giuliani used to belittle garden supporters, giving cover to the city's efforts, as if gardens and housing were zero-sum instead of mutually supporting solutions. The problem of framing was vexing.[25] City council member Carlina Rivera failed to engage supporters of the consensus plan, while the mayor's office framed the conflict as a mere communication problem. "Rivera and her allies posited a strict binary choice between protecting the safety and well-being of NYCHA residents and protecting trees and the park's diverse ecosystem," noted John Tarleton.[26] Of course, this approach obscured issues with flooding due to rain, excess heat, the toxicity of synthetic turf, and the lack of information made available to engineers who did the review. Some movements are better than others at framing a problem, ideally with a diagnosis of a condition, in a way that serves as a call to arms, mobilizing and engaging the problem. Most of us wanted a plan that included natural systems and social justice; did not use materials and methods that exacerbate climate change; supported the well-being of the community, park, and environment equitably; and provided clear storm protection. The mayor wanted a different kind of redesign for the park, one that did not inconvenience motorists.[27] ESCR only addresses storm surge and sea-level rise, but not heat, rainfall, and other health-impacting elements of climate change. The city said their plan was stronger at protecting the community immediately adjacent to the park, and they had the *New York Times* to echo their case. The gap between the community and politicians was vast.

We ended the year with a dance party along the East River Park, dancing with friends from the community well into the night.

In early January 2022, the activist group 1000people1000trees put out a call: "TODAY, come out!!! Gather 8AM inside #EastRiverPark Houston Entrance." There, activists faced the January cold and climbed some of the trees in East River Park, where they hoped for a stay to save them from demolition.

"We love you," several of us greeted the activists sitting in the branches and looking out at the river while the gusting wind blew off the water into the trees.

Savitri D. posted a message after getting home: "Greeting the dawn in the arms of a great sentinel. Thank you Earth. Honor and protect the trees, and the life they support, known and unknown." Still, the pace was wearing on everyone. It was hard to keep up with the daily, ongoing assault on the trees.

"I love you cherry trees," said Eileen Myles, that February. "You are a family . . . it's sooo great to see you . . . here. . . . I look up and you are in the sky . . . you are wiser than me. Thank you trees."

Myles wrote, "The plan generally bugged the city because they said it was too much work."[28] The mayor contended that the ESCR was the best "engineering solution" for addressing the issue of flooding. The engineering—with its cement structures, entrenched kin-in-construction contracts, and status as partners with (or handmaiden to) developers and media powerhouses overriding smarter, more sustainable, socially just models of adaptation in discourse and landscape— emerged as another countervailing force with which to contend. What they were not saying was that their plan offered double the construction contracts without closing lanes of traffic on the six-lane FDR Drive, noted Tarleton.[29] "The community was involved," Myles went on. "They were asked what they wanted. People said they wanted the park to remain, they wanted to see the water, they wanted flood protection."[30]

Whenever I could during the spring, I rode to East River Park to visit the cherry trees the city was cutting down. Billy, Sylver, Eileen, and Sarah were there to slow the machines.

"No ecocide on the Lower East Side," declared activists, blocking the trucks moving in, blocking, locking themselves, and taking busts as the trees and cherry blossoms were taken out. The baron construction site was the picture of something like an apocalypse.

"The field so empty, / The trees so still, / Wondering where did all their bodies go?" said Sara Matthes, reading a poem at one of the poetry gatherings at the park.

Kathryn Freed, who was representing the activists in a lawsuit against the city, posted a commentary on the debacle: "The city's destruction of more than half of East River Park in the name of resiliency . . . continues to be the source of contention, anger and even fear in the Lower East Side. . . . Certainly, having workers shout 'Happy Earth Day!' while they butchered the cherry trees didn't help."[31] The cross-cultural antipathy between the construction crews and environmentalists was ongoing.

In the summer of 2022, I arrived at 6th Street grove, one of the last stands of trees remaining, where the Earth Church was "meeting at East

River Park to say goodbye to the grove of trees before they're cut down." Bill Talen, in his guise as Reverend Billy, a preacher turned environmental performance artist, was preaching to the trees. Some seven hundred had been killed, with more in the coming months.

"Listen to the trees," preached Reverend Billy. "As a part of a flood mitigation plan the City of New York will level the park and build a concrete flood wall, killing more than one thousand mature trees in the process," he lamented, wondering if we've been entertaining ourselves to death. "Central Park, even Union Square has a conservancy, with backers such as Michael Bloomberg. No such conservancy is available for East River Park, just a low-income constituency. I fear luxury towers will be going there next," said Billy, echoing Naomi Klein's argument about shock doctrine. "And now this monoculture is coming in, not like a forest. Five hundred trees were cut, but five hundred are still standing, half one story, half another story," he concluded.

Reverend Billy alluded to a larger pattern of governance in New York since the fiscal crisis of the 1970s. "Though private coffers renovated major parks, this contributed to a two-tiered park system bifurcated by race and class," contends historian Benjamin Holtzman.[32] In the renovation of East River Park, this dynamic is vexing; the park was renovated, with more asphalt, and access to the roadway was, and remains, unimpeded. The poor are left with less green space; the rich keep their central parks and vacation homes. It's a pattern that repeats itself again and again as gaps in access to green space expand along with cleavages over access to resources. Income inequalities are often expressed through spatial inequalities, in access to public space in general and green space, in particular here. This trend toward privatization favors business-friendly opinions and approaches, compelling rebuilding bigger and better, over and over again. The case of East River Park speaks to the importance of access to public space and the disconcerting results of having a metropolitan landscape so shaped by class and race.[33]

Conclusion: What Can We Learn from This Case?

A highway runs through the universe in Douglas Adams's novel *The Hitchhiker's Guide to the Galaxy*, not unlike Robert Moses's highway through the South Bronx that destroyed neighborhoods. Place matters.[34]

It gives us meaning. It mattered when Moses bulldozed the South Bronx. It mattered in the battle for East River Park.[35]

"One year since #Eastriverpark began losing its majestic trees," Wendy Brawer posted. "When will we stop destroying ecosystems in the name of resilience? #ESCR is a model of what not to do."

What happened in East River Park? With a city ever in flux, the dynamics are many, as are the construction projects in Lower Manhattan, with neighborhood members and public space users creating community by the water ever clashing with newer, more affluent tower inhabitants hoping for the expensive water view but never stepping outside except when getting into cars. Water is an actor; waterfront access is an ongoing struggle.

What can we learn from this conflict? What does it tell us about the fragility of riverfront parks and the relationship of marginalized community groups with the city? What is the relationship of community public space needs and the new movement to protect parks and the city from rising water levels, especially during storms? Throughout this chapter, I attempted to answer these questions and a few others. To do so, I traced a story about climate change and urban ecology, using reflective ethnography to consider a few competing narratives of urban space, exploring the ways organizers used poetry and direct action to frame and guide their actions, reinforce values, and stir emotions while assessing a complicated, albeit flawed, campaign.[36]

It's no understatement to suggest the city and the community had a difficult time hearing each other when the city rejected the consensus plan in favor of the ESCR. One of the most significant lessons is about city priorities and the lack of communication or engagement with involved communities. It's never easy for two sides to hear one another, but it can be done. This was the story of the Dudley Street Neighborhood Initiative a generation ago in Boston. The group cleaned up their neighborhood, but only after beating back a top-down plan from city hall and fashioning a consensus plan with city planners to revitalize their community.[37] The same dynamic did not take place in East River Park, with its competing claims and cultural conflicts and a consensus plan that was discarded rather than amended by the city.

David Snow and Robert Benford argue that effective framing of a mobilization campaign requires effective diagnosis of a problem, a proposed solution to that problem that resonates, and a call to arms for mobilizing support to implement that change.[38] Throughout the struggle for East River Park, organizers struggled with framing. Instead of the successful call to "save the gardens" that worked in the 1990s, they framed their campaign in more complicated terms: "Save our park," "Save the trees," and "Fight climate change." And the mobilization never gained the traction to win over enough supporters or beat back the one-dimensional call to "halt flooding." All the while, the city simply said "We can save the park" (by destroying the trees), splitting the community with arguments about who really are the stakeholders and who are not.[39] "Rivera and her allies didn't create the neighborhood's racial divide, but they did exploit them for everything they were worth," said Tarleton.[40] But it's not like there was robust grassroots support for the current plan, either. Pro-ESCR rallies were small and mostly made up of staffers from elected officials' offices and allied nonprofits. More people signed petitions against ESCR than voted for Rivera in her first primary. There was more diversity in opposition to the plan than coverage often suggested, yet the framing was mixed.[41] Democracy can be unwieldy in complicated situations such as this.

Still, climate change is real. Solutions to what we are doing about it are anything but clear. (Well, solutions are clear but there's political opposition to them.) Many argue that for more biodiversity, via more trees, more green space is an important tool to combat climate change.[42]

It is incumbent upon policymakers to bridge the gap between communities and politicians, serving as mediators and translators between elected officials and community members. The consensus plan lacked political backing. Still, had Rivera voted the other way, we might have had the consensus plan, which was what experts designed and what the community wanted. The gap between the administration and the community was a fault line. Top-down solutions do not work for the community. Place matters; urban waterfronts matter as spaces that illuminate, even as the commons is threatened in cities, beaches, and waterfronts around the world.

NOTES

1 Rebuild by Design, *The Big "U"* (New York: BIG Bjarke Ingels Group, n.d.).
2 "East Side Coastal Resiliency Project," Institute for Sustainable Infrastructure, August 1, 2022, https://sustainableinfrastructure.org.
3 John Tarleton, "Carlina Rivera and the Untold History of How East River Park Was Destroyed," *The Indypendent*, August 23, 2022, https://indypendent.

org/2022/08/carlina-rivera-and-the-untold-history-of-how-east-river-park-was-destroyed/.

4 K. Krajick, "New York's Waterways Are Swimming in Plastic Microbeads," *State of the Planet*, August 16, 2017, https://news.climate.columbia.edu.

5 David A. Snow and Robert D. Benford, "Ideology, Frame Resonance and Participant Mobilization," in *From Structure to Action: Comparing Social Movement Across Cultures*, ed. Bert Klandermans, Hanspeter Kriesi, Sidney Tarrow, 197–218 (Greenwich, CT: Jai Press, 1988); Alexa Trumpy, "Based on a True Story: The Use of Conversion Stories in Social Movements," *Social Movement Studies* 21, no. 5 (2022): 642–58.

6 Myles Zhang, "The Privatization of Public Space in Lower Manhattan," Myles Zhang (blog), April 20, 2021, www.myleszhang.org.

7 Defend the Atlanta Forest, "Atlanta-Area Park Carelessly Destroyed: Bulldozer & Cops Escorted Out of the Atlanta Forest," May 9, 2022, https://defendtheatlantaforest.org.

8 Steve Butters, "The Logic-of-Enquiry of Participant Observation," in *Resistance through Rituals: Youth Subcultures in Post-War Britain*, ed. Stuart Hall and Tony Jefferson (London: Hutchinson University Library, 1983), 253–73.

9 Carolyn Ellis and Art Bochner, "Autoethnography, Personal Narrative, Reflexivity: Researcher as Subject," in *The Handbook of Qualitative Research*, 2nd ed., ed. Norman K. Denzin and Yvonna S. Lincoln (Thousand Oaks, CA: SAGE, 2000), 733–68. After all, US historian Howard Zinn (1994) reminds us, there is no neutral participation. See Howard Zinn, *You Can't Be Neutral on a Moving Train: a Personal History of Our Times* (Boston: Beacon Press, 1994).

10 Eugénie Birch, "Cities, People, and Processes as Planning Case Studies" in *The Oxford Handbook of Urban Planning*, ed. Rachel Weber and Randall Crane (Oxford: Oxford University Press, 2012), 259–84.

11 Michaela Benson and Karen O'Reilly, "Reflexive Practice in Live Sociology: Lessons from Researching Brexit in the Lives of British Citizens Living in the EU-27," *Qualitative Research* 22, no. 2 (2020): 1–17.

12 Jim Carroll, *The Basketball Diaries* (New York: Penguin Books, 1987), 14–17.

13 Eileen Myles, "Eileen Myles Chronicles a People's History of East River Park," *Document*, July 5, 2022, www.documentjournal.com.

14 City of New York, "Sandy and Its Impacts," in *A Stronger, More Resilient New York* (n.p.: City of New York, 2013). https://www.nyc.gov.

15 Arbor Day Foundation, "Trees Help Fight Climate Change," www.arborday.org; United Nations, "Biodiversity—Our Strongest Natural Defense against Climate Change," www.un.org, accessed August 23, 2024.

16 Tarleton, "Carlina Rivera and the Untold History."

17 Benjamin Holtzman, *The Long Recovery: New York City and the Path to Neoliberalism* (New York: Oxford University Press, 2021).

18 Naomi Klein, *The Shock Doctrine: The Rise of Disaster Capitalism* (New York: Henry Holt, 2007).

19 This theme is highlighted in chapter 6 of this volume by Iveson and Vila-Concejo.

20 Village Sun, "Justin Brannan Now Ducking East River Park Activists on Coastal-Resiliency Plan," *Village Sun*, October 19, 2021, https://thevillagesun.com.

21 "We're fighting this with poetry, politically, and the courts," says Eileen Myles. "This was part of what I loved about it when I moved here in the 1970's." I read Allen Ginsberg's "Tears" (1956), thinking about the trees. Allen Ginsberg, "Tears," 1956, *andtheblossom* (blog), https://andtheblossom.wordpress.com. Countless poems grew from this space. See Alex Vadukul, "Eileen Myles Watches Over an Ever-Changing New York," *New York Times*, May 18, 2022, www.nytimes.com.

22 Eileen Myles, "Mayor Robber: Eileen Myles on the Pointless Demolition of Manhattan's East River Park," *Artforum*, May 17, 2021, www.artforum.com.

23 Sydney Pereira, "Local Groups Sue City Over Plan to Overhaul East River Park," *Gothamist*, February 7, 2020, https://gothamist.com.

24 Michael Kimmelman, "What Does It Mean to Save a Neighborhood?," *New York Times*, December 2, 2021, www.nytimes.com.

25 Snow and Benford, "Ideology, Frame Resonance and Participant Mobilization"; Trumpy, "Based on a True Story.

26 Tarleton, "Carlina Rivera and the Untold History."

27 Tarleton.

28 Myles, "Eileen Myles Chronicles a People's History of East River Park."

29 Tarleton, "Carlina Rivera and the Untold History."

30 Myles, "Eileen Myles Chronicles a People's History of East River Park."

31 Kathryn Freed, "Opinion: City Breaks Promises on East River Park as Con Ed Fells Trees in Northern Half, Air-Monitoring Reports Withheld," *Village Sun*, May 12, 2022.

32 Benjamin Holtzman, *The Long Crisis: New York City and the Path to Neoliberalism* (New York: Oxford University Press, 2021).

33 Holtzman, *Long Recovery*.

34 John Dixon and Kevin Durrheim "Displacing Place-Identity: A Discursive Approach to Locating Self and Other," *British Journal of Social Psychology* 39, no. 1 (2010): 27–44.

35 For many of us, the actions to save the park constituted forms of what Della Porta describes as "eventful protest." These are moments that acknowledge "the power of events in history." See Donatella Della Porta, "Eventful Protest, Global Conflicts," *Distinktion: Journal of Social Theory* 9, no. 2 (2008): 27–56.

36 Trumpy, "Based on a True Story."

37 Leah Mahan and Mark Lipton, *Holding Ground: The Rebirth of Dudley Street*, documentary 00:58:00, 1996, https://www.newday.com/films/holding-ground-the-rebirth-of-dudley-street.

38 Snow and Benford, "Ideology, Frame Resonance and Participant Mobilization."

39 Kimmelman, "What Does It Mean to Save a Neighborhood?"

40 Tarleton, "Carlina Rivera and the Untold History."

41 Tarleton.

42 Alan Buis, "Examining the Viability of Planting Trees to Help Mitigate Climate Change," NASA, November 7, 2019. See also United Nations, "Biodiversity."

PART III

Racializing the Beach

Inequality and Erasure

Sammy's Beach (Setha Low, 2023)

9

Resilience in Rockaway

Coastal Imaginaries and Racial Coastal Formation

BRYCE DUBOIS AND LEIGH GRAHAM

Superstorm Sandy (hereafter, "Sandy") struck the Rockaway peninsula (a.k.a., "Rockaway") in Queens County, New York City, on October 29, 2012, causing fourteen-foot storm surges and severe flooding from the Atlantic Ocean and Jamaica Bay.[1] Widespread physical damage from the peninsula's colloquial "west end" to "east end" included power outages in public housing, burned homes in Belle Harbor and Breezy Point, and the erasure of more than 1.3 million cubic yards of sand and destruction of over two miles of Rockaway's beloved boardwalk (see figure 9.1). At Community Board 14 (CB14), Rockaway's influential resident advisory group that operates as part of NYC's urban planning process, discussions of how to restore the beach and boardwalk dominated meetings for at least eighteen months after Sandy, despite Jamaica Bay's floodwaters also inundating bayfront properties and Black activists from the east end of the peninsula calling for a focus beyond the beach.

We, at the time a Doctoral Candidate in Environmental Psychology and an Assistant Professor of Urban Affairs both interested in the politics of public space and redevelopment in periods of crisis, were drawn to this impassioned public discussion about uneven coastal restoration and community recovery priorities and embarked on research to understand residents' experiences with disaster recovery processes and potential coastal impacts. We embarked on intensive fieldwork conducted from January 2013 through June 2015, using participant observation, in-depth interviews, and documentary and archival research. During that period, we interviewed fifty-six residents (forty) and practitioners (sixteen) from across the peninsula, including leaders of nongovernmental organizations and affiliates engaged in Rockaway Beach stewardship,

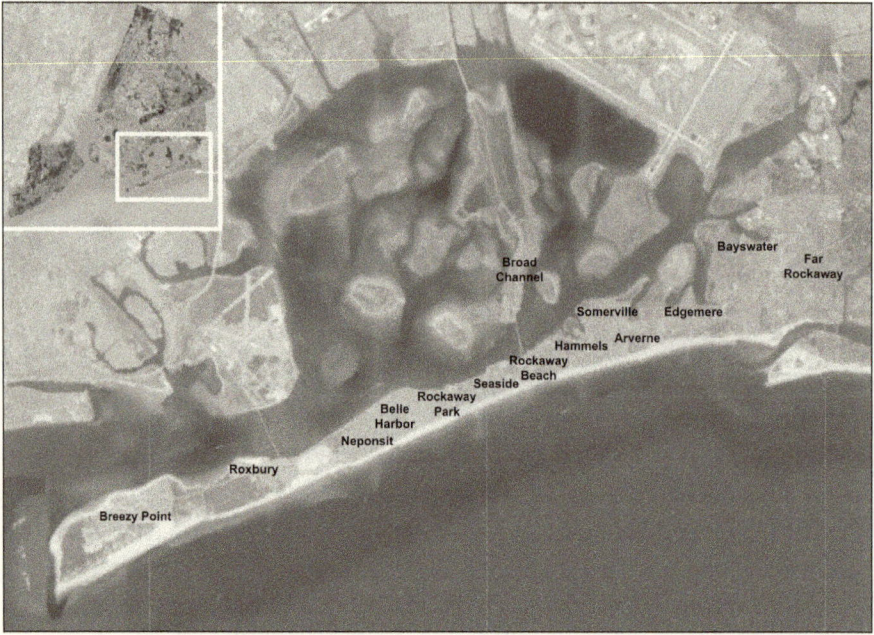

Figure 9.1. Neighborhoods of the Rockaway Peninsula. Source: City of New York, OpenStreetMap contributors (2023), https://www.openstreetmap.org/.

civically active residents, NYC Housing Authority tenant association leaders, local bureaucrats, and elected officials. Utilizing a descriptive coding approach and drawing on Wolford's analytical concept of spatial imaginaries,[2] we present the coastal imaginaries of local civic actors— shared conceptions of life on the coast and how they seek to live—to understand their engagement with state climate adaptation plans after Sandy. In doing so, we show that these adaptation plans are racial projects in this case study of racial coastal formation in Rockaway.[3]

Racial Coastal Formation

R. Dean Hardy, Richard Milligan, and Nik Heynen developed the concept of racial coastal formation to center uneven racial development in research on the "climate gap," the chasm between scientific data and "everyday concerns of vulnerable communities" in climate policymaking.[4] Racial coastal formation extends from Michael Omi and Howard

Winant's seminal theory of racial formation, in which race is a "socio-historical concept" made real categorically and meaningfully through "social relationships and historical context."[5] When racial formation happens through policymaking, the state's responsibility for racial subjugation in society is revealed.[6] Racial coastal formation intentionally and explicitly incorporates the historical legacies and contemporary burdens of systemic racism with state policymaking and practice that leave coastal communities of color distinctly vulnerable to sea-level rise and other impacts of climate change. Hardy, Milligan, and Heynen offer this as a corrective to "colorblind adaptation," which they critique as ignoring structural racism or systematically attributing its impacts to another cause (e.g., poverty).

Consider this frequent twentieth-century legacy along the US coast: a city places public housing in inexpensive, low-lying areas at a city's watery edge, later expands access for low-income Black households to this deeply subsidized yet socio-spatially isolated housing, and later must make do with strenuous federal cuts to operating and capital budgets, leaving these households particularly exposed to twenty-first-century climate risks as governments at all levels fail to act in response to climate change. In this scenario, colorblind adaptation attempts to address climate risks without attention to the specifics of this legacy that left low-income Black families in harm's way or what that suggests for interventions. In contrast, a framework of racial coastal formation deliberately takes into account violent legacies of racial capitalism, such as living in a disinvested community first wrought from urban renewal that rendered Black lives "un-geographic,"[7] now downstream of a new market rate development whose stormwater features redirect water overflow to pool in this subsidized development down the block. This coastal infrastructure becomes a repeating hydrogeological formation that intensifies the negative valence of socio-racial space as climate change adaptation unfolds.[8]

A Brief Socio-history of Rockaway

In the late seventeenth century, plantation-settler Richard Cornell brought ten enslaved people to the peninsula and supplanted the Indigenous Lenape-Canarsee peoples from their summer fishing grounds in what is now known as Far Rockaway.[9] The peninsula stayed sparsely inhabited until about 1833 when the majority of land was sold by Cornell's descendants to the Rockaway Association. Commercial and residential development ensued on bulkheaded and "improved land," aided by the extension of the Long Island Railroad in the 1880s. Irish immigration to New York City began with the earliest waves in the 1840s due to the Great Famine, and Irish Americans traveled to labor in Rockaway in the late 1800s. Later, beginning in the 1910s, African Americans, many of them coming from the south during the Great Migration, came to work in the hotels and summer businesses. Housing for summer residents and year-round workers included tent cities and bungalows in low marshy areas.

Racial covenants and anti-Black sentiment forced African Americans into three neighborhood pockets: Hammels, between the railroad tracks and Rockaway Beach Boulevard; Redfern, at Far Rockaway's county line; and a small zone in Arverne.[10] Within these neighborhoods, Black-led organizations advocated and provided for public health, economic opportunity, and housing for residents in the face of disinvestment.

This activism challenged a range of local racial projects, especially Robert Moses's "Rockaway Improvement" urban renewal plan.[11] This plan incorporated Rockaway Beach and boardwalk into the NYC Parks system, developed the shorefront, and supported beach maintenance and beach ecology. In the process, Moses systematically displaced thousands of predominantly Black residents, first from Redfern (1949–52), then Hammels (1949–54), and finally, Arverne (1964–1972).[12] Some residents were forcibly moved all three times. Moses's collaborator, the Rockaway Chamber of Commerce, was described in 1961 by the Colored Democratic Association and the New York City Commission on Intergroup Relations as an "anti-negro establishment."[13] In contrast, Moses eventually capitulated to the organizing of Broad Channel and Breezy Point, Irish American neighborhoods that were also targeted for removal.

Rockaway development in the twentieth century included extensive public housing, with all six complexes placed on the bay side. Community District 14 is in the top ten districts for public housing (21 percent of rental units) and has the highest concentration of public housing in Queens. Public housing residents in Rockaway are predominantly Black and compose 8 percent of the peninsula's population. Racial and economic stratification is further evident across Rockaway's three neighborhood areas, per 2013 census data tabulated by the NYC Department of City Planning: Far Rockaway (population: 50,314) and Arverne (36,255), both on the east end, and the west end (29,464). Far Rockaway has a plurality of Black residents, 26 percent white residents, and 26 percent Hispanic residents. Seventy-one percent of residents in Far Rockaway rent, mostly in multifamily properties. Approximately 20 percent of families live in poverty. Arverne, in the center of the peninsula, is predominantly Black and has five of Rockaway's six public housing developments (the sixth is in Far Rockaway). The west end is majority white (78 percent) single-family homeowners (66 percent). At 35 percent, the west end has one of the highest concentrations of Irish Americans in the United States.[14]

Today, the city owns the majority of the Atlantic Ocean coastline in Rockaway, as compared to a mix of public and private ownership along Jamaica Bay's shoreline. Rockaway Beach is one of the most expensive beaches in America to maintain due to erosion. Since 1926, the federal Army Corps of Engineers (USACE) has renourished the beach thirty-seven times, for a total cost of $313,456,885 converted for inflation in 2021 dollars.[15]

Post-Sandy Response: The Special Initiative for Rebuilding and Resiliency

One month after Sandy, Mayor Mike Bloomberg committed to rebuilding the Rockaway Beach boardwalk, with details to come.[16] Two weeks later, Bloomberg unveiled the $20 billion Special Initiative for Rebuilding and Resiliency (SIRR), which committed to rebuild the city's waterfront—its "sixth borough"—using "smarter and stronger and more sustainable" measures.[17] The "greatest risk" for NYC's coastal communities was storm surge; sea-level rise was considered a minor to moderate risk (through 2050). Public input from Rockaway on SIRR emphasized coastal protection on the ocean and bay sides and local job training and creation.

One SIRR initiative was "Get New Yorkers 'Back to the Beach' for summer 2013"; by winter 2012, the Parks Department and USACE began repurposing more than 100,000 cubic yards of sand as a temporary barrier against future storms. The former Parks Department administrator for Rockaway Beach told us that "city hall was right here, and the elected officials. Everybody was really, 'What can we do?' . . . There was a real rush to create things . . . decisions were being made citywide. . . . We are part of a much bigger plan, to operate all the beaches. Some things were done really quickly, and miraculously." Over seven months—with crews working around the clock for four—the city spent more than $140 million to restore the beach, an "engine of commerce" and "symbol" of the city bouncing back.[18] USACE pumped in 600,000 cubic yards of sand during summer 2013.

In September 2013, the city announced its plan to rebuild the board-walk using concrete to serve as the primary infrastructure to protect residents from storm surge along the oceanfront. Successfully securing $480 million from the Federal Emergency Management Agency, the city rebuilt the boardwalk in four phases through summer 2017, including an extra two feet of sand above USACE designs. Beach restoration activities were implemented starting in the west and slowly moved east, mirroring the uneven sociopolitical context of the peninsula; the east end finally received its first beach nourishment in fall 2022, just ahead of the tenth anniversary of Sandy.[19]

Bifurcated Coastal Imaginaries

In the period following Superstorm Sandy, we found that coastal imagi-naries were bifurcated in Rockaway between the east and west ends according to whether SIRR-led beach restoration responded to com-munities' needs. These bifurcated imaginaries map onto racial and class stratification on the peninsula. For example, an east end nonprofit direc-tor's remarks describing local politics were echoed in several interviews: "There's Rockaway west and east. . . . They've never got along, but they never really needed a reason to get along. There was nothing we ever worked on. [Now long-term recovery groups are] still working on their bylaws because there's just so much conflict between all the residents of different groups, and [they're] trying to figure out who should have the power." Here, the economic development leader implicitly marks the race and class antipathy present in Rockaway. While the Irish American com-munity of the west end was able to prevent displacement tactics from the city in the mid-twentieth century, they had since organized against the prevalence of public, subsidized, and single-room occupancy on the peninsula. But whereas the classist and veiled racist framing of the peninsula as a "dumping ground" was commonly invoked in CB14 deliberations by west end civic actors, east end civic actors had been striving for housing security and economic development, including for the residents of the more than four thousand units of public housing who had been this target of west end ire. East end residents described that the benefits of the budding "hipster economy" of the beach failed to extend sufficiently eastward pre-Sandy and that official bodies like CB14 and other formalized planning groups failed to attend to this post-Sandy. Below, we present how this sociopolitical context was articulated through east and west end coastal imaginaries in post-Sandy adaptation planning.

West End Coastal Imaginary

BEACH RESTORATION TO MAINTAIN GENERATIONAL COMMUNITY CONTINUITY

Rockaway was long known as NYC's "Irish Riviera."[20] In our interviews, a narrative from white west end residents was lovingly told of a family-centered, close-knit way of life, one organized around membership in the various Catholic parishes and connected ancestrally and symbolically to Ireland, linking these maritime communities. As one interviewee told us, "My grandfather was from Sligo and if you've ever . . . seen Sligo, Ireland, the wave break is like Rockaway." Another said, "Rockaway is a beach community, and it's a very close-knit community, so you develop long term relationships with the people in Rockaway. And they don't go, they stay, generation after generation after generation. So you really have roots here, in Rockaway." Among this community, interviewees described a "beach-oriented" life shared by a "circle" of friends and kin tending to children who "grew up on the beach. . . . [They would] be down there every day." "It's the only way to bring up children," a multigenerational family patriarch told us, comparing the safety of children moving about on a wide-open beach landscape to a swimming pool's slippery, hard-surfaced dangers. In these imaginaries, the boardwalk was the physical and psychological spine bridging neighborhoods, generations, and local histories. As one west end activist told us, "The boardwalk was amazing . . . you would just hang on the boardwalk and that interconnectivity at the boardwalk, you might have known someone because you went to St. Rose [Parish], you may have known a kid because you played basketball or your cousin went to St. Rose [Parish] or even to my house." Among west end Rockaway residents, people are labeled by their length of tenure and relationship to the beach. Visitors to Rockaway Beach are called DFD, "Down for the Day." This includes the Brooklyn-reared spouse of one Irish Catholic Rockaway resident who moved to Rockaway in 1975 after they got married. He told us, "My wife [is] from Rockway but I'm a DFDer. . . . Down for the day, because I've only been here thirty-five, forty years."

Sandy severely ruptured this imaginary; very quickly, the coastal protection required to ensure that west enders' desires to age in place and bring up families in this seaside way of life became the focus of discussion. As explained by the cochair of the Parks Subcommittee of CB14, a sympathetic Latino with a rare oceanfront home on the east end, in this "shore community . . . for some of us, we lost our identity. The reason why so many of us are out here . . . it's not just the beach, it's a space to be on the beach, the boardwalk . . . that became the focal point, I think because it was the one thing that I think people could really affect change on." This example is just one of many that we heard where restoration of the beach and boardwalk was intertwined with discussions of identity and sense of place.

WEST END COMMUNITY SUFFICIENTLY PROTECTED THROUGH BEACH INFRASTRUCTURE

This cohesive place attachment enabled residents to collaboratively learn about and advocate for various environmental protection features (e.g., trap bags, rock jetties [groins], baffle walls, Hesco barriers) directly to city and federal planners.[21] Their post-Sandy activism built on prestorm knowledge of environmental vulnerability led by a voluble minority of west end homeowners demanding protection for their homes *and* protection for their beach as a treasured asset. This group, now calling themselves Rockaway United after Sandy, was repeatedly cited in media interviews as having "warned" the city about the risks of living alongside a heavily eroded coastline. Members were highly visible at CB14 and other planning meetings concerning beach restoration.

Vocal west end homeowners tended to consider the beach and boardwalk as an extension of their private property. Excessive sand, wood, and debris from the beach and boardwalk dispersed up west end streets and in homeowners' yards. The primary emphasis in interviews was on rebuilding homes and using coastal infrastructure such as rock groins and sand nourishment to protect private property, that "protection is a number one" goal. "Without the proper protection, you're always in a precarious situation, depending on the weather," one couple told us. Although they were uncertain about whether another "Sandy" might happen again, they felt that even the possibility of it meant they had to "fight so hard for the right protection."

Elected officials, CB14 appointees, and planners embraced this framing, including the responsibility for community learning and decision-making. The staff of NYC council member Eric Ulrich, who represented the west end, described protection as a "very big" priority post-Sandy: "I think once the majority of people got into their homes, they're all like, 'Okay, how do we stop this from happening again?' So we're learning all about coastal erosion protection, the difference between groins, jetties, and breakwaters, and all the little things that come with it so we can try and explain it to people." The parks subcommittee cochair felt the boardwalk dominated public discussion because residents view it as coastal "protection." "People were clamoring at the Parks Department, 'When are you going to provide protection?'" He recalled that the Parks Department made explicit that their mandate was not to provide protection, that fortifying the beach is the responsibility of the USACE, but he believed residents "needed a scapegoat only because of what happened with the boardwalk to begin with, the fact that the boardwalk was not moored to the pilings. . . . The boardwalk came off, it damaged some people's houses, and the people lost, I guess, because we all lived by the ocean we all used the boardwalk." The New York State Assembly member who represented most of Rockaway's white-majority neighborhoods in the lower house of the New York State legislature, including Far Rockaway's Orthodox Jewish community, broke down the protection needs of the various neighborhoods. Wealthy, oceanfront enclaves Belle Harbor and Neponsit on the west end wanted "their baffle wall" to protect them from the sea. Rockaway Park and Rockaway Beach, on the west end where most of the public swimming beaches are, needed "sand dunes or [to] rebuild the beaches." His office got involved in the boardwalk planning process to be "deliberate" and ensure that "every community concern was heard, every piece of community input was implemented and the final product is gonna [sic] be something that not only is aesthetically pleasing . . . but will also serve as a protection. . . . So when the next storm comes and it will come, the boardwalk will not just stand up and stay as a boardwalk but it will act as prevention for community damage and that is critical and that is key." He elaborated on the need to rebuild a resilient boardwalk: "I was born and raised in Rockaway and I love this community but I'm raising my kids here and it's not about me. It's that when they grow up and they get married, I want them to stay here, too; if we don't do it right, they're not gonna want to." Coastal protection, in the form of a concrete boardwalk and sand dunes, became the central focus of a broader range of designs that residents hoped would protect and preserve the west end's community. Moreover, Rockaway United's protection frame expanded to include shielding community identity from outside influence, such as developers targeting the growing hipster enclave in the Beach 90s in the Rockaway Beach neighborhood on the west end. They made a literal connection between their neighborhoods and the beach by describing how sand replenishment was an intervention to protect and stabilize their majority Irish American middle-class community. Local desire for influence over beach restoration planning arose because of its relationship to their private property and because of west enders' inextricable experience of the benefits of the public beach as meeting their personal and private needs.

Officially justified by the state of damage to the public swimming beaches and the vulnerability of nearby homes to oceanside flooding, the city's five-year Rockaway Beach Restoration Project began on the west end in 2014, less than two years after Sandy. This is despite the fact that while these interventions were designed to buffer storm surge and wave impacts, they could not prevent flooding from waves that washed over the boardwalk and dunes, nor protect from bayside flooding and a rising water table that cause nuisance (tidal) and other forms of flooding that are unimpeded by the beach protection.

East End Coastal Imaginary

NEED FOR WHOLE COMMUNITY ADAPTATION

On the east end, the beach did not feature prominently in coastal imaginaries, its restoration deemed insufficient to meet poststorm calls for jobs and economic opportunities that amplified pre-Sandy concerns. An activist for east end equitable redevelopment summarized the priority issues in boardwalk reconstruction and Sandy recovery plans: "It's been very clear across the board that jobs is number one . . . local hiring and training for those jobs and careers is number one. There is a big need for jobs out here. Then after that would be affordability so making sure that a good percentage of the housing is available to people who live in the Rockaways because our area median income is significantly less than the whole city." A highly regarded faith-based leader described "protecting our community" as the number one priority, followed by "rebuilding our community" through short- and long-term economic opportunities on the east end. The east end's NYC council representative further prioritized local hiring via boardwalk reconstruction in post-Sandy efforts, building on the momentum of opening a Workforce One Center in Far Rockaway right before Sandy. Despite this coherence concerning the lack of economic opportunity on Rockaway's east end, the area experienced multiple barriers to influencing public sector priorities.

A WHOLE BARRIER BEACH ECOSYSTEM PERSPECTIVE IMPEDED BY JURISDICTIONAL CONUNDRUMS

Like their west end neighbors, east end residents have a strong attachment to life at the ocean's edge. African American, Caribbean, and Latino residents talked about its allure: the "cool breeze" in the summertime, the fresh scent of ocean air, the evocation of ties to childhood by the Virginia coast. As one homeowner put it, the ocean is "a very powerful magnet. I would never live anywhere else if I could help it." Yet, east end residents lived with much greater exposure to Jamaica Bay and its potential floodwaters than west end residents. As the SIRR attests, a primary obstacle to addressing Jamaica Bay protection is its complex management and ownership. As one east end civic leader told us about the bay, "It's federal, state, and city that own different portions, plus private. That's what makes the bay so hard to fix. Everybody has a little piece of the bay."

Yet, east end residents also felt that debates about sand and coastal protection at CB14 obscured discussions of other needs. One African American public housing resident, then one of the city's youngest community board members, summed up CB14: "I had to understand, what is [the] community board? Why am I coming to [the] community board? I would ask this when I was younger, because I've been with the [redacted community organization] since I was sixteen. You know, 'Why do we come to these three-hour meetings once a month, and sit here until 10:30 at night?' To hear people talk about sand."[22] For CB14, she said, "one of the biggest pressing issues [is] coastal protection, especially on the western end of the peninsula." Another public housing resident described a similar disconnect with CB14 in different terms: CB14 "is not for public housing, that's not where their interest lies. . . . It's basically about homeowners, so there's really no interest in there for us to really work with them. . . . We feel like we don't fit in." One of CB14's elected officers, an African American and an east end renter in Bayswater, described her experience bringing up bayside vulnerability:

> One of the things that people found annoying, from me, was why a lot of people want to talk about the beach, the beach, the beach, and what happened on the west end. And I don't take away that what happened on that end was devastating. But guess what, on the east end we also have Jamaica Bay. . . . They got devastated too. But nobody talks about that. . . . That's why they're sick of me. Because I don't let them forget about the bay. . . . I try to bring it up every time I can. . . . I think there's a little prejudice . . . when it comes to talking about coastal protection . . . and what you get done, and whose area is going to get done first.

The municipal emphasis on the beach and boardwalk via the SIRR as key to the peninsula's recovery limited east end residents' ability to have their concerns about economic development and bayside vulnerability addressed. This coastal protection prejudice prevailed both spatially and temporally. Near the second anniversary of Sandy, NYC council member Donovan Richards was still calling for repairs to bulkheads and other damaged infrastructure in historically Black Arverne on the east end's bayside.[23] During this same period, the city repaired the baywall in affluent Belle Harbor on the west end, and budgeted approximately $22 million to raise streets, improve drainage, and install bulkheads in Broad Channel, a heavily Irish Catholic community in Jamaica Bay that is part of CB14.[24] As of 2017, the proposed $3.6 billion "system of flood walls, levees and gates to control water levels in Jamaica Bay" by the USACE lay dormant in Congress and "there have been virtually no improvements to the bayside since the storm," reported the *New York Times*.[25]

Discussion

All our interviewees expressed a commitment to life in Rockaway. Yet coastal imaginaries of life in Rockaway from the minority-majority east end to the white-majority west end diverged in alignment with government priorities for Sandy recovery. The systematic nature of the inequity of the Rockaway coast produced through contemporary state projects that resemble urban renewal era priorities (i.e., the priorities of white beachfront neighborhoods dovetail with state projects while Black neighborhoods with greater bayfront risk are sidelined in planning processes) illustrates the racialized logics of constructing coastal space in climate adaptation efforts.[26]

Both east and west end residents have coastal imaginaries that were physically and psychologically impacted by Sandy and the state projects that followed. For west end stakeholders, the SIRR emphasis on the beach as the primary state resilience project restored their coastal imaginaries. Their ability to cogovern in this process was built on generations of white supremacist policy decisions, at least since redlining in the contemporary period, and afforded them continued property and generational protection and eventual increases in housing value.[27] Conversely, east end stakeholders viewed the SIRR emphasis on the beach as insufficient to their needs. Indeed, Black coastal imaginaries in Rockaway tell the story of organized self-determination to produce and maintain community in the face of disinvestment and dispossession.[28] This legacy underpins post-Sandy advocacy such as Rockaway Wildfire's calls for a holistic, barrier beach ecology and cooperative development.

Inattention to bayside protection, employment, and housing security for Black residents are examples of the state's continued practice of nullifying Black spatial belonging in Rockaway.[29] Historically blocked access to prime coastal waterfront increases African American communities' protection from storm surge in the twenty-first century. Yet, their state-sponsored settlement in low-lying marginal land along the bay exposes them to the risk of sea-level rise while also making them an afterthought to inundation risks. Thus, these climate exposures are *racialized*, due to colorblind adaptation decisions now that build on policy legacies of enforced residential patterns from at least redlining. The SIRR encapsulates colorblind adaptation and its disproportionate reliance on technocratic data, actuarial experts, and climate scientific analysis rather than experiential, everyday life, place attachment, material from communities as is called for by climate justice advocates.[30] (The SIRR also never used the word "race" as a demographic characteristic in one of the most racially diverse cities in the United States.) The commodified public beach created conditions where this economically valuable area was prioritized at the expense of vulnerable communities, producing what Isabelle Anguelovski and colleagues describe as an "act of omission" in climate adaptation.[31]

The SIRR leveraged a concept of urban resilience that maintained inequity, prioritizing infrastructure investment for the city's economic assets in a way that upheld and reinforced existing racial disparities.[32] This led to the rather incongruent outcome of a dramatic increase in real estate sales activity in areas hit by Sandy,[33] including new high-end developments that have increased home values in the historically low-income communities of Arverne and Far Rockaway.[34] Enshrining inequity through resilience establishes the conditions for resilience gentrification post-Sandy, as seen elsewhere in NYC,[35] with double dispossession for vulnerable residents from both climate change *and* gentrification.[36]

Conclusion

This case study of racial coastal formation in Rockaway after Sandy demonstrates how resilience becomes part of a state project commodifying public space through beach restoration, creating conditions for resilience gentrification that benefit white neighborhoods, and increasing climate and displacement risk for Black residents. While this study did not include participation in more recent planning efforts, such as the Downtown Far Rockaway Redevelopment Plan, and likely amplifies prominent voices through use of a snowball sampling, the sociohistorical context and discourses from east and west end stakeholders makes clear that the SIRR continued racialized state projects. Specifically, the SIRR hampered east end efforts for economic security, community well-being, and coastal protection while creating opportunities for west end stakeholders to coproduce Rockaway's climate adaptation efforts in a way that strengthened their psychological and material ties to the peninsula. In doing so, Rockaway Beach restoration after Sandy exemplifies the process of racial coastal formation. We hope this case informs climate praxis in its illustration of how climate adaptation efforts reproduce radically uneven rights to transform and benefit from these processes if such efforts fail to attend to racialized geographies of coastal space.

NOTES

1 Eric Blake, Todd Kimberlain, Robert Berg, John Cangialosi, and John Beven II, "Tropical Cyclone Report Hurricane Sandy (AL182012)," National Hurricane Center, October 22–29, 2012.

2 Johnny Saldaña, "The Coding Manual for Qualitative Researchers," *Coding Manual for Qualitative Researchers* (Thousand Oaks, CA: SAGE, 2021); Wendy Wolford, "This Land Is Ours Now: Spatial Imaginaries and the Struggle for Land in Brazil," *Annals of the Association of American Geographers* 94, no. 2 (2004): 409–24.

3 R. Dean Hardy, Richard A. Milligan, and Nik Heynen, "Racial Coastal Formation: The Environmental Injustice of Colorblind Adaptation Planning for Sea-level Rise," *Geoforum* 87 (2017): 62–72.

4 Hardy, Milligan, and Heynen, "Racial Coastal Formation," 62–63, quoting Jean-Christophe Gaillard, "The climate gap." *Climate and Development* 4, no. 4 (2012): 261-264.

5 Michael Omi and Howard Winant, *Racial Formation in the United States* (New York: Routledge, 2014), 15.

6 Kevin Fox Gotham, "Urban Space, Restrictive Covenants and the Origins of Racial Residential Segregation in a US city, 1900–50," *International Journal of Urban and Regional Research* 24, no. 3 (2000): 616–33.

7 Katherine McKittrick, *Demonic Grounds: Black Women and the Cartographies of Struggle* (Minneapolis: University of Minnesota Press, 2006).

8 Isabelle Anguelovski, Linda Shi, Eric Chu, Daniel Gallagher, Kian Goh, Zachary Lamb, Kara Reeve, and Hannah Teicher, "Equity Impacts of Urban Land Use Planning for Climate Adaptation: Critical Perspectives from the Global North and South," *Journal of Planning Education and Research* 36, no. 3 (2016): 333–48.

9 Andrew Bellot, *History of the Rockaways from the Year 1685 to 1917: Being a Complete Record* (Far Rockaway: Self-published, 1922).

10 Lawrence Kaplan and Carol P. Kaplan, *Between Ocean and City: The Transformation of Rockaway, New York* (New York: Columbia University Press, 2003).

11 Moses, Robert. "The Improvement of Coney Island, Rockaway, and South Beaches," *New York City, NY: Department of Parks*, 1937.

12 Kaplan and Kaplan, *Between Ocean and City*, 119.

13 Kaplan and Kaplan, 109.

14 Jed Kolko, "America's Most Irish Towns," *Huffpost*, www.huffpost.com.

15 Program for the Study of Developed Shorelines at Western Carolina University, "U.S. Totals," Beach Nourishment Viewer, accessed September 4, 2022, https://beachnourishment.wcu.edu.

16 Liz Robbins, "Mayor Says Rebuilt Boardwalk in Rockaways Won't Be Wooden," *New York Times*, November 30, 2012, www.nytimes.com.

17 Michael Grynbaum and David Chen, "Bloomberg Puts Soothing Aside as He Rushes to Bring Back City," *New York Times*, November 15, 2012, www.nytimes.com.

18 Lisa Foderaro, "Racing the Clock in the Rockaways," *New York Times*, May 18, 2013, www.nytimes.com.

19 Bobby Calvan, "Superstorm Sandy Legacy: Recovery Far from Equal on NY Shore," *US News*, October 26, 2022, www.usnews.com.

20 Corey Kilgannon, "Irish Toast a Summer Paradise Lost in the Rockaways," *New York Times*, September 25, 2007, www.nytimes.com.

21 Lisa Colangelo, "New Rockaway Boardwalk to Loom Large Over the Beach," *New York Daily News*, October 9, 2013, www.nydailynews.com.

22 See also Leigh Graham, "Public Housing Participation in Superstorm Sandy Recovery: Living in a Differentiated State in Rockaway, Queens," *Urban Affairs Review* 56, no. 1 (2020): 289–324.

23 Lisa Colangelo, "Struggling Rockaway Hopes $274 Million Boardwalk Project Will Yield Local Jobs," *New York Daily News*, March 25, 2014, https://www.nydailynews.com.

24 Kia Gregory, "Where Streets Flood with the Tide, a Debate Over City Aid," *New York Times*, July 9, 2013, https://www.nytimes.com.

25 Luis Ferré-Sadurní, "Could the Rockaways Survive Another Sandy?," *New York Times*, July 13, 2017, https://www.nytimes.com.

26 Hardy, Milligan, and Heynen, "Racial Coastal Formation," 62–72.

27 Pablo Herreros-Cantis, Veronica Olivotto, Zbigniew J. Grabowski, and Timon McPhearson, "Shifting Landscapes of Coastal Flood Risk: Environmental (In)justice of Urban Change, Sea Level Rise, and Differential Vulnerability in New York City," *Urban transformations* 2, no. 1 (2020): 1–28.

28 Ruth Wilson Gilmore, "Abolition Geography and the Problem of Innocence," *Tabula Rasa* 28 (2018): 57–77.

29 McKittrick, *Demonic Grounds*.

30 Mike Hulme, "Meet the Humanities," *Nature Climate Change* 1, no. 4 (2011): 177–79.; Julie Koppel Maldonado, "A Multiple Knowledge Approach for Adaptation to Environmental Change: Lessons Learned from Coastal Louisiana's Tribal Communities," *Journal of Political Ecology* 21, no. 1 (2014): 61–82; Anguelovski et

al., "Equity Impacts," 333–48; Jennifer L. Rice, Brian J. Burke, and Nik Heynen, "Knowing climate change, embodying climate praxis: Experiential knowledge in Southern Appalachia," *Annals of the Association of American Geographers* 105, no. 2 (2015): 253–62.

31 Anguelovski et al., "Equity Impacts," 333–48.

32 Stephanie Wakefield, "Urban Resilience as Critique: Problematizing Infrastructure in Post-Sandy New York City," *Political Geography* 79 (2020): 102148.

33 Mike Quintana, "Hurricane Sandy NYC: City Builds Where Floods Hit Hardest." *StreetEasy* (blog), October 25, 2017, https://streeteasy.com; Emily Erdos, "Hurricane Sandy and the inequalities of resilience in New York: The American Prospect," *American Prospect*, October 5, 2018, https://prospect.org.

34 Herreros-Cantis et al., "Shifting Landscapes," 1–28.

35 Kenneth A. Gould and Tammy L. Lewis. "From Green Gentrification to Resilience Gentrification: An Example from Brooklyn," in *Taking Chances on the Coast After Hurricane Sandy*, ed. Karen M. O'Neill and Daniel J. Van Abs, 145–63 (New Brunswick, NJ: Rutgers University Press, 2018).

36 Dean Hardy, Maurice Bailey, and Nik Heynen. "'We're Still Here': An Abolition Ecology Blockade of Double Dispossession of Gullah/Geechee Land," *Annals of the American Association of Geographers* 112, no. 3 (2022): 867–76.

10

Only in Albuquerque

Remembering Albuquerque's Tingley Beach

NATASHA HOWARD

Albuquerque, New Mexico, once had a beach! What makes this so surprising is that New Mexico is a landlocked state. However, in the 1930s, the City of Albuquerque created the Conservancy Beach just south of the downtown area by diverting water from the Rio Grande into a series of small ponds. The new recreational area was advertised and promoted as Conservancy Beach. The beach had several names over the couple of decades that it existed. It was Conservancy Beach, Ernie Pyle Beach—named after the Pulitzer Prize winner who lived in the city for a short period of time—and later, Tingley Beach. Local people and the city newspapers often referred to it as Tingley Beach in honor of famed New Mexico politician Clyde Tingley. Though Clyde Tingley is credited with creating Albuquerque's beach, prominent environmentalist Aldo Leopold is thought to have first proposed a similar idea to city officials somewhat earlier, with Tingley expanding on that original idea. For two decades, Tingley Beach was one of the most popular recreational spots for Albuquerque's residents.

During its heyday, Tingley Beach was an important recreational space with a sandy shoreline for sunbathing and public swimming along with a large indoor bathhouse for soaking (figure 10.1). Tingley Beach held boat races, diving contests, beauty pageants, and it had lifeguards and offered swim lessons. The beach is known to have been wildly popular with locals and visitors alike. The *Albuquerque Progress* magazine reported in 1949 that the beach was one of the most popular recreational spots in the entire city. In the 1950s, Tingley Beach was closed to swimmers due to a fear of polio. It was transformed into a fishing pond and nature area, which continues to survive today as part of the larger Albuquerque Biological Park (ABQ BioPark).

What remains as insight into its heyday as a beach can be found in a wide collection of photos and a black-and-white film that are archived at the Albuquerque Museum, the Albuquerque Public Library Special Collections, and on the City of Albuquerque's website. Images of the public beach are part of a permanent exhibition at the Albuquerque Museum titled *Only in Albuquerque*. Visitors to the City of Albuquerque's website and the Albuquerque Museum's exhibit can learn about the historic importance of this little beach in the middle of the desert. The photographs highlight a space where many people gathered, swam, and socialized. The socializing aspect was paramount because the beach allowed Albuquerque residents to play in the sun and fraternize with other Albuquerqueans. The photos certainly display the social aspect of the beach. Well-known city leaders—including political figures like Clyde Tingley—spent time at the beach interacting with residents. Therefore, the beach represented more than a place to simply cool off; it was indeed *the* cool place to see and be seen. The photographs attest to this. The photographs were also used to market Albuquerque to potential new residents as an inviting oasis in the middle of a hot desert.

The first photo I came across was interesting because it captured a beach in the middle of landlocked Albuquerque. It is also an area I had visited and driven often, but I had no idea it was previously a swimming beach. Though the area always had the name Tingley Beach, I knew (like all Albuquerqueans) that it was not, in fact, a swimming beach now. I always thought the name was tongue-in-cheek. The area was actually a place where people hiked along the Bosque, went on bike rides, fed the hundreds of ducks, or did some small-scale fishing in the little pond. I had never known it to look like what the photos depicted. As I did more research, I discovered there were about a hundred photos of the beach, archived primarily in the Albuquerque Museum. The Albuquerque Museum has a permanent exhibit that includes photos and a short history of the beach. The beach is presented in a way that points to Albuquerque's early history of advocacy for public recreational leisure spaces and conservation of the city's natural environment. Despite the public presentation of the beach as one that all the city's residents enjoyed, I knew the city's nature spaces had their own racial politics. I also thought there was no way to understand that history without examining the photos in relation to other events that were happening to Black residents in Albuquerque and elsewhere in New Mexico.

Figure 10.1. Tingley Beach 1930s to 1940s.
Courtesy of the Albuquerque Museum
Photo Archives.

In this chapter, I will argue that how the beach has been remembered teaches important lessons about race and space in Albuquerque, New Mexico. The legacy of how the beach represented racialized space is present in photo archives, a museum exhibit, and a documentary film on Albuquerque's Open Space System. The photos are a visual history of the racialization of space at Tingley Beach. One of the more interesting things about this beach—beyond the fact that a swimming beach existed in the desert—is that despite the rich archive of photos, film, and other ephemera about the beach, there seems to be no documentation of Black visitors to Tingley Beach, even though there was a Black section of town very close to the area. Albuquerque's Black residents are remarkably absent from any waterscape depictions in the city, and their erasure is a central theme of this chapter.

I am interested in how the photos depict the racialization of beach space. What do the photos depict about race, recreational beach space, and a white spatial imaginary? What do the photos say about how powerful people wanted the city represented and publicized to people living within the state and those living beyond? How do the photos align with other popular racialized imagery about New Mexico as a space that, while multicultural, is still a white space for white enjoyment?

Methodology

Sites of memory are socially constructed.[1] Memories are carefully selected, and some are diminished or erased over time. Robyn Autry notes the importance of examining the politics of representation and collective memory. Autry highlights how claims to representation (i.e., spatial claims) have to do with establishing belonging. Signaling who belongs, however, is deeply rooted in determining who has material rights, including land and water resources. While Autry's work unpacks the politics surrounding museums as "sites of memory," I examine how a historic public beach, which no longer exists, is today memorialized as a site of memory through a museum exhibition, as well as through other visual and written histories documenting the origins of the beach. How do public institutions work to create a collective memory with the intention of influencing how the viewer *feels* about a particular place? What are the politics of this kind of memory work? Museums and city websites have a political purpose beyond simply providing "factual" information about a place at a particular point in time; rather, what these kinds of historic exhibits do is represent part of the spatial politics of a place. We must question what challenges are presented when partial stories are represented as collective experiences, a collective memory, and therefore, a collective truth. Whose stories have been omitted, and what, if any, are the consequences of these kinds of omissions? More importantly, how would a more complicated story make viewers feel about that historic space? Would the nostalgic feelings these kinds of exhibits tend to foster be challenged in any way?

I draw upon techniques found in visual culture analysis and examine a photo exhibition on Tingley Beach.[2] Additionally, I analyze written discourse about Tingley Beach that has been presented on the city's website and other documents articulating the history of the beach. I highlight how whiteness is deployed through the photos and nostalgic remembering of the history of the beach. My concern is how the beach has been historicized as a white spatial imaginary through a visual interpretation that clearly marks the historic beach as having been white space.[3] Even though the photos demonstrate that Hispanics likely visited the beach as well, their presence does not challenge the idea that Tingley Beach was white space. In fact, their presence may have been one of the ways that some members of Albuquerque's Hispanic population pushed the boundaries of whiteness by being included in Anglo recreational space.

The Politics of Memory and Memorializing Places

Steven Hoelscher notes that "the importance of memory to geographers is magnified by its acknowledged social nature; the political and material consequences of white are evident everywhere."[4] Similarly, Derek Alderman, Jordan Brasher and Owen Dwyer III argue that "memorials and monuments are important symbolic conduits for not just expressing certain versions of history but casting legitimacy upon them. They give the past a tangible and familiarity, making the history they commemorate appear to be part of the natural, taken-for-granted order of things. Memorials and monuments influence how people remember and *value* the past, in part because of their apparent permanence and the common impression that they are impartial recorders of history."[5] While the focus here is not necessarily on monuments, the photos do that same work of commemorating and honoring a particular time in Albuquerque's history. The images and accompanying narratives are presented to the public as a way of valuing a historical moment. The City of Albuquerque has made an important effort to remember its beach history, not just because it is a cute story about a beach in the desert but because it is also nostalgic for remembering a white spatial past. The photos of the beach are similar to how Route 66 has been memorialized in Albuquerque as a wonderfully free leisure place to travel and play. And yet, many places along that route in Albuquerque refused to serve African Americans or allow them to stay in the famous hotels.[6] Still, Route 66 is fondly imagined as a place of leisure and enjoyment where everyone "got their kicks." Yet it is important to remember that not everyone got their kicks on Route 66. For some, Route 66 was not welcoming. Similarly, historian Robyn Autry reminds us that institutional efforts to retell the past are always about creating a shared sense of the past. In the case of Albuquerque's Open Space System, the history is made public through signs at open space sites, the city's parks and recreation website, year-round public events at open spaces, and middle and high school curricula. How Albuquerque came to have such a network of beautiful open spaces is told without any attention given to the racial politics behind the city's open space history.

Albuquerque's Open Space System includes more than 1,100 acres of urban wildland and wildlife spaces, mountain trails, river and bosque environments, agricultural land, and cultural monuments (e.g., the Petroglyph National Park), as well as the Albuquerque volcanoes and canyon hiking areas. Collectively, they form an extensive network throughout the city called the Open Space System. Tingley Beach is credited as one of the earliest endeavors in creating that system.[7]

Museum exhibits and city websites are places where people learn about who made important spatial contributions and who could enjoy those places. Photos of Tingley Beach speak to the value of a little water oasis in the desert for recreation and leisure; they also speak to *who* was valued and, therefore, able to take advantage of this manmade oasis. Though there is no official record of the beach being legally marked as a whites-only space, the photos offer evidence that perhaps Black people did not feel comfortable visiting the beach. The fact that the surrounding neighborhoods were, in fact, legally designated as spaces where

Black people were not welcome likely had consequences for how Black people felt about traveling to the beach and enjoying it. In other words, we have to consider the wider scale repercussions of racialized housing covenants that effectively said this is a no-Black-people zone.

Clyde Tingley and Albuquerque's Open Space Movement

The City of Albuquerque's website and Albuquerque Museum note the role of Clyde Tingley and his efforts to have the beach constructed. Tingley was a powerhouse in New Mexico's history. He was famously charismatic and had a politically powerful social network—including a well-known friendship with Franklin D. Roosevelt. He held multiple political offices in the state, including as an alderman in 1916–17; a member of the city commission in 1922; chair of the city commission (the equivalent of mayor) in 1925–35, 1939–46, 1947–48, and 1951–54; and New Mexico's state governor in 1935–39. Tingley is remembered as having championed outdoor leisure and recreational spaces drawing upon the natural resources of the city—most notably the Rio Grande. Tingley strongly advocated for marketing the city to tourists as a place to enjoy recreation and leisure in nature. The idea was to have a natural oasis in a rapidly growing urban environment that would welcome visitors and residents. Under Tingley's administration, parks were built throughout the city. He famously put a significant amount of the city's resources into construction along the river to create a swimming beach. The *Albuquerque Journal* routinely reported on each major step in building the beach. The journal printed photos of Tingley with the workers and engineers who made his dream of a city beach a reality. As I came across more and more photos and a series of black-and-white film clips from the era, I was curious as to why there were no images of Black people at this beach. I knew that the historic Black community of South Broadway was not far from the beach. I carefully searched all the photos and other documented information about Tingley Beach but could not find any evidence that Black people ever enjoyed the beach like white residents did.

Black Residents under Tingley's Leadership

Under Tingley's political leadership, Albuquerque's Black residents had a different experience with space. Black residents experienced confinement and restrictions on their movements. A survey taken by city officials in 1951 reported that most of the city's residents believed that Black people felt the brunt of discrimination more systematically than other groups in Albuquerque—signifying anti-Black discrimination was coded in everyday practices. A key detail highlighted in this report is that anti-Black spatial discrimination had a long and consistent record in the city. The South Broadway neighborhood was becoming the primary place for Black residents because of anti-Black housing covenants that began appearing in all neighborhoods. These covenants prevented Black people from owning or renting homes in most neighborhoods in Albuquerque.

South Broadway, located just east of the downtown area, near the railroad tracks, was the only place Black people were permitted to live. Even before the housing covenants legally appeared, testimony from a white builder noted that there had always been a gentleman's agreement to not sell or rent property to Black people.[8] Racially discriminatory policies like housing covenants that barred Albuquerque's Black residents from entire neighborhoods had other consequences, like making it difficult for Black people to enjoy public recreational resources that should have been available for all. The Black community faced significant de jure and de facto segregation and racial discrimination under Tingley's leadership. The photo archive has no images of signs forbidding Black people from entering Tingley Beach, yet there is no indication in them that Black people ever swam at or visited the beach. How, then, can we think of Tingley Beach as white space?

Elise C. Boddie theorizes that spaces, like people, have a racial identity (as well as how spaces are racialized) and that this racialization of space can determine who is allowed to be in those places comfortably. Defining racial territoriality, she states, "Places can have a racial identity and meaning based on socially engrained racial biases regarding the people who inhabit, frequent, or are associated with particular places and racialized cultural norms of spatial belonging and exclusion. This racial meaning has consequences that constitutional law often overlooks. One consequence is 'racial territoriality,' a distinctive form of discrimination in which people of color are excluded from public spaces that are identified as 'white' and treated as being only for white people."[9] In the case of the city's Black residents, they were confronted with spatial policies and practices that effectively restricted their free movement in the city—and therefore, their ability to enjoy public recreational spaces was also impacted. The beach was, by default, white space because it was an extension of white residential neighborhoods that were not open to Black people. The logic of racial covenants applied to home ownership also had implications for public spaces in or near those racially exclusive neighborhoods. Anti-Black racial ideology was part of the discriminatory atmosphere of the larger city space.

Black residents were in heated battles to access amenities like restaurants, hot springs, hotels, and unrestricted housing.[10] This did not only affect residents, it also racially segregated the tourism market. Recreation and leisure happened within Black residents' backyards, highlighting what George Lipsitz says about the Black spatial imaginary.[11] Black homes have often been the spaces for community leisure and recreation, even for Black tourists.[12] And while the resiliency of Black communities is celebrated for creating Black leisure spaces, we have to acknowledge that this resilience emerged as resistance to white supremacist, anti-Black racism, which limited Black people's access to recreational leisure spaces. This would surely have an effect on how Black people viewed their right to enjoy a space like Tingley Beach.

Institutionalizing a Collective Memory

City websites can provide educators, visitors, and local residents with important information about spatial history. Similarly, museums are important educational spaces that do more than just teach facts; they also work to inform or stimulate the viewer's sentiments and feelings about place and space—even if they were not there. Exhibits can stir up feelings or emotions about particular events or periods of time in our history. By visualizing what the city looked like in a particular era, viewers can imagine how they might have experienced recreation, had they lived in that era. The viewer can also ascertain what might have been spatialized social dynamics at a particular point in time—in this case, the era of the 1930s to 1950s in Albuquerque, New Mexico. Those who lived through that period may think nostalgically about what Tingley Beach represented for its residents. On one social media page, I found people fondly remembering swimming at the beach and people talking about what a wonderful place it was to visit. The historic photos are a reminder of the kind of recreational space city residents enjoyed. Except, not all city residents enjoyed the space equally.

Speaking about the spatialization of race, George Lipsitz notes that by focusing exclusively on Black disadvantage, we may unintentionally conceal how Black disadvantage is linked to white privilege and white advantage.[13] In this chapter, I have connected white spatial advantage with Black disadvantage to demonstrate how they are intimately related. Historic photos demonstrate how the space was specifically marked as racially white space. How the beach is remembered through public exhibitions, websites, and other educational materials also reminds us that the beach is, even today, historicized and remembered as a white space. Black Albuquerqueans did not enjoy similar kinds of recreational space.

Not only are there no photos of Black people visiting Tingly Beach, but there are no historic photo images of Black people enjoying any of the city's recreational nature spaces in a similar fashion. By this, I mean that you will not find a collection of Black people enjoying nature in the same way as Albuquerque's white population does. There is no comparable compilation of photos pointing to Black residents taking up space in the same way as demonstrated in the Tingley photo collection. What does this say about Black people's belonging and right to recreational space? What does this say about whiteness as the ability to take up space in a way that non-white people are unable to do?

Taking up space means more than just your physical presence. By this, I also mean that you feel like you belong in the space. You feel like you have the right to adjust the space to meet your needs. You feel that other people can and will accommodate your presence. You ultimately feel that you have the right to be there. Just as white residents and visitors were taking up space, Black residents and visitors were being cordoned off in the tiny South Broadway neighborhood. They were subjected to increasing amounts of surveillance.

As the city expanded under the leadership of Clyde Tingley, each new neighborhood adopted anti-Black housing covenants. Expansion and development became equivalent with white actors attempting to protect "their" space from Black people entering. Businesses consistently denied Black people service.[14] Schools were integrated, but Black students were grouped and segregated within the classroom.[15] Furthermore, by confining Black residents to one neighborhood, there was a ripple effect of de facto segregation. There is no reason to believe this did not extend to recreational spaces like Tingley Beach. In other words, if white people expected social advantages and exclusively white spaces in housing, schooling, restaurants, hotels, and other businesses, it would not be unusual for them to expect exclusively white recreational space as well. Whiteness comes with expectations.[16] This includes spatial expectations, which have historically meant that public spaces could become exclusive for white people—what has essentially been the privatization of public spaces for white people's use and enjoyment. Historically, a city's public recreational spaces align with a white spatial imaginary. The history of nature spaces is that they have frequently been officially or unofficially designated as white spaces, meaning that people of color are imagined as not belonging in that space and not possessing any rights to be there.[17] In the case of Tingley Beach, Black residents were dispossessed from the enjoyment of land and water.

However, I want to return to the larger issue of how this is remembered today, given there is such an effort to memorialize Tingley Beach and remind city residents about this space. Why today is there such nostalgia for the tiny human-made beach? Why the efforts to curate a memory about the historic beach? The displays have sought to create a collective memory that does not align with the historical spatial experiences and memories of Black people in the city. New Mexico has long established itself as a multicultural mecca without the kind of harsh anti-Black racism found elsewhere in the country. De jure segregation was random, or so we are to believe. Efforts to memorialize Tingley Beach align with the "soft" discrimination discourse. Soft discrimination hides the violence of the white spatial imaginary. It paints the white spatial imaginary in warm, nonthreatening, sepia tones that did not really have significant consequences for Black people. The exhibits are designed to foster a feeling of warmth and nostalgia for a time when Albuquerque was in its heyday. It was a developing place that offered beautiful landscapes and recreational enjoyment in the middle of the desert.[18] And yet, despite all the beautiful natural resources available, only certain people seemed to routinely enjoy them and feel they had a right to enjoy the space unbothered—those whose race did not dictate where they could and could not go or how they would be received when they arrived.

The exhibit *Only in Albuquerque* underscores how the city is unique. Photos of Tingley Beach are situated in that exhibit under the theme of "resourcefulness" to demonstrate how insightful the city was for turning a swampy area along the Rio Grande into a beautiful beach that residents could enjoy during the hot desert summers so common in Albuquerque. On the other hand, much of the city's history looks remarkably similar to other parts of the United States. This is particularly true when we examine anti-Black spatial discrimination, which disrupts the "unique" multicultural tolerance discourse that city officials too often advertise.

The historic photos of Tingley Beach depict a natural landscape with a racially white connection to the land. Can they then also be part of the colonial imagination? The pictures do the work of reproducing racial ideology. Given how they are exhibited at the museum and discussed on the city's website, they reinscribe New Mexico's tricultural myth. This myth would uphold the idea that Anglo, Hispanic, and Native American people are the rightful heirs to the land of New Mexico—because these groups have made important contributions to the state's history and, therefore, are worthy of being memorialized. (I have discussed the tricultural myth of New Mexico in more depth elsewhere.[19]) Though this myth has been contested and challenged many times, the myth is routinely perpetuated in visual representations of the state. Triculturalism has its own hierarchy as well. Anglos are presented as the pinnacle of the tricultural mythology. That is to say, Anglo settlers were supposedly the ones who brought science and rationality to New Mexico. Not so ironically, this is depicted in the exhibit featuring Tingley Beach. The section of the exhibit that depicts Tingley Beach is characterized as "resourceful" and imagines Tingley Beach as the vision of an ingenious idea of an Anglo settler (Clyde Tingley but also Aldo Leopold) for creating something as imaginative as a beach in Albuquerque.

The idea that there is one collective memory has to be interrupted. Museum spaces provide important history lessons for visitors. Middle and high school teachers use these spaces to educate their students about "truth" and historical "fact." We have to strive for more complicated lessons that disrupt the idea that there is a collective spatial memory. We can present to visitors the complicated racialized spatial history that has been differentiated by race and space. The tribute to Tingley Beach is a visual reminder at odds with the colorblind language accompanying it. Colorblind rhetoric used in the exhibit emphasizes collectively through the use of words like "our," "we," and "us." On the other hand, the photo images are a stark reminder that this recreational space was racially exclusive for white Albuquerqueans.

The colorblind discourse promoting the exhibit speaks to the ways whiteness works. The ideology of whiteness works first to mask the tangible privileges associated with it and then make it appear innocent to the onlooker. The photos of white people playing in water, sunbathing, boating, and swimming are also connected to how white people controlled public spaces in Albuquerque. By focusing on the novelty of a beach in a desert and the images of play and leisure outdoors, we may forget that the exhibit also teaches about whiteness as a spatial structure. In other words, these playful historical images are also lessons on the spatialization of whiteness in Albuquerque.

New Mexico is also a place where the past is very important to how people view the spatial environment. Reliving and reimagining the past is important to the state's identity. The New Mexico tourism industry has always promoted to the visitor an opportunity to relive the past. It is what makes places like Albuquerque appealing to so many. The right to exclusively use and enjoy public spaces is part of the inheritance of whiteness. The ability to create and maintain exclusively white public spaces symbolizes white people's natural right to the land and, in this case, the shoreline too.

NOTES

1 Robin Autry, *Desegregating the Past: The Public Life of Memory in the United States and South Africa* (New York: Columbia University Press, 2017). Citing French historian Pierre Nora, Autry argues that often museums are "sites of memory" or places where memory "crystallizes and secretes itself" (18).

2 Philip J. Deloria and Alexander Olson, *American Studies: A User's Guide* (Oakland, California: University of California Press, 2017); bell hooks, *Black Looks: Race and Representation* (Boston, MA: South End Press, 1992).

3 George Lipsitz, *How Racism Takes Place* (Philadelphia, PA: Temple University Press, 2011).

4   Steven Hoelscher, "Making Place, Making Race: Performances of Whiteness in the Jim Crow South," *Annals of American Geographers* 93, no. 3 (2003), 657–86.

5   Derek Alderman, Jordan P. Brasher, and Owen Dwyer III. "Memorials and Monuments," *International Encyclopedia of Human Geography*, 2nd ed., vol. 9, ed. Audrey Kobayashi, 39–47 (Amsterdam: Elsevier, 2020).

6   Candacy Taylor, "The Roots of Route 66," *Atlantic Magazine*, November 3, 2016.

7   City of Albuquerque, "Open Space History," Open Space One Albuquerque, accessed November 1, 2023, www.cabq.gov/. See also documentary on Albuquerque's Open Spaces https://www.youtube.com/@OneABQMedia/videos (accessed December 2023).

8   "Colored Citizens of New Mexico Sore Over Jim Crow Code," *Southwest Review Newspaper*, March 21, 1925.

9   Elise Boddie, "Racial Territoriality," *UCLA Law Review* 58, no. 2 (2010): 401–62.

10 Roger Banks, "Between the Tracks and the Freeway: African Americans in Albuquerque," in *African Americans in Albuquerque*, ed. Bruce Glasrud (Albuquerque: University of New Mexico Press, 2013), 170–85; George Long, "How Albuquerque Got Its Civil Rights Ordinance," in *African Americans in Albuquerque*, ed. Bruce Glasrud (Albuquerque: University of New Mexico Press, 2013), 165–69.

11 George Lipsitz, *How Racism Takes Place* (Philadelphia: South End Press, 2011).

12 This point was raised in an oral history interview collected by the author. The participant was an African American woman who was raised in Albuquerque. The participant noted that when she was young, the backyard of her parents' home served as an entertainment and socialization place for Black residents and travelers. Her family often put up travelers in their home.

13 George Lipsitz, "The Racialization of Space and the Spatialization of Race: Theorizing the Hidden Architecture of Landscape," *Landscape Journal* 26, no. 1 (2007): 10–23.

14 Long, "How Albuquerque Got Its Civil Rights Ordinance."

15 "African Ameircan Rights," MyText, accessed December 10, 2023, https://mytext.cnm.edu. Note that African American students were separated from the rest of their graduating class.

16 Cheryl Harris, "Whiteness as Property," in *Critical Race Theory: The Key Writings That Formed the Movement*, ed. Kimberle Crenshaw, Neil Gotanda, Gary Peller, and Kendall Thomas (New York: The New Press, 1995), 276–91.

17 Christabel Devadoss, "Monolith, the Colonial Face of Landscape Photography," *Gender, Place and Culture: A Journal of Feminist Geography* 29, no. 1 (2022): 26–51; Carolyn Finney, *Black Faces, White Spaces: Reimagining the Relationship of African Americans to the Great Outdoors* (Chapel Hill: University of North Carolina Press, 2014).

18 Erna Fergusson, *New Mexico: A Pageant of Three Peoples* (New York: Alfred A. Knopf, 1951).

19 Natasha Howard, "The Promise and Purpose of Black Studies in New Mexico: Spatializing Blackness in New Mexico," *Africology: The Journal of Pan African Studies* 12, no. 9 (March 2019): 33–40.

11

Four Shades of Whiteness

A History of the White South African Beachgoer

KEVIN DURRHEIM

This chapter recounts a story about settler whiteness. It is set in South Africa and focuses on the thin slice of social life connected to vacationing and beaches. This might seem peripheral to the business of life at work, home, and school, but, as Ana Deumert argues, the "beach is central to understanding how settler-colonial spatialities (and affects) are instantiated in public space."[1] Their status as public space makes beaches an ideal site to investigate racial entitlement. Place entitlement is a sense of rights to and ownership of territory you don't own but nonetheless occupy and possess. Colonialism was an expression of racial entitlement and the beaches in the postcolony exemplify the enactment of racial entitlement to public space.

We look to the beaches to see how settler whiteness has been enacted and defended in South Africa. Whiteness is strongly tied to *privilege* in the popular and political imagination, an invisible knapsack of unearned privileges that bestows untold advantages on white-classified persons and secures persistent systemic inequality.[2] As beaches became spaces of vacationing and leisure, they formed a symbolic and material anchor for white privilege. However, political struggle and anti-racism victories have left whiteness in a state of "postcolonial melancholia,"[3] bearing a sense of loss of untrammeled privilege and exclusion. This pervasive condition has also brought into view a second defining element of whiteness, namely, an egocentric sense of *entitlement*.[4] This is a sense of worthiness tied up with entitlement to property and privilege overlaid with sense of loss, rage, and shame when expected benefits or advantages are denied.[5]

We begin centuries after settlers arrived in their ships, disembarked onto foreign shores where they met the Native inhabitants and proceeded to occupy and own the whole of the land. Their entitlement was so thoroughgoing, their displacement of the native inhabitants so unrelenting, that settler communities around the world came to see themselves as natives![6]

We take up the story in the 1960s, the postwar period of industrial expansion and capital accumulation in South Africa, and proceed through three further historical periods of the development of whiteness. Each period describes an expression of entitlement that together represent the four shades of whiteness of the title. The first two periods, the white playground and disruption, are premised on the de jure segregation of the apartheid period. The second two, displacement and counterattack, are postapartheid expressions. The shifting sands of whiteness show how privilege and entitlement are reworked in changing social and political contexts and go some way to explain its stubborn persistence.

The White Playground

The first shade of whiteness develops in the political context that John Cell described as the "highest stage of white supremacy."[7] This is the context of de jure segregation, where settlers establish political control over Natives in a context of urbanization, modernization, and industrialization. The Native Urban Areas Act (1923) and later the Group Areas Act (1950 and 1966) legislated racially demarcated residences in South African cities. Apartheid legislation also prohibited mixed marriages, segregated schools and employment, and reserved advanced curricula and professional categories of work for white people.[8] The hope to create a grand white homeland also required the regulation of public space. To this end, the 1953 Reservation of Separate Amenities Act sought to minimize contact between races by racially demarcating public space, creating "whites-only" versus "nonwhite" park benches, public toilets, railway station entrances, beaches, and so on.[9] The best beaches were designated white by law, force, and intimidation.

The transformation of beaches into "vacationscapes" played a major role in their racialization.[10] Beaches at Durban and along the Natal Coast accommodated over a quarter of domestic tourists between 1985 and 2000.[11] Up to the early 1980s, the upcountry holidaymakers were exclusively white. The racial segregation and inequality of meant that before the 1970s, there was "virtually no Black tourism market."[12] Durban allocated the safest, best resourced, and most central 58 percent of beaches to white people, who constituted 22 percent of the population at the time.[13] Beaches allocated to Black, Indian, and other people of color were outside of municipal control and ill-suited to recreational use.

The material order of the highest stage of white supremacy was overlaid with a symbolic order of whiteness. In the early twentieth century, Durban was marketed as "the Brighton of South Africa," self-consciously modeled on the British beaches with promenade, pier, "palatial hotels," and safe swimming.[14] With growing Afrikaner Nationalism in the postwar period, the Britishness of Durban gave way to a "Californication" of South African beach culture.[15] This was modeled on the "political economy of race and racism in the beach communities of Southern California and the discursive structures that grew up around it—the cult of California surfside leisure, . . . youth culture, and the intertwined mythos of freedom . . . and unfettered heteronormative sexuality."[16] Durban was reconfigured as Surf City, "a symbolic marker for racial exclusivity on the apartheid beach."[17] The racial exclusiveness of surfing, sunbathing, and holidaymaking and the representations of (white) healthy, youthful selves and happy families all provided markers of whiteness and racial privilege.

Claudia Plunkett conducted a comprehensive search of pictures of people on the beach published in the *Natal Mercury* between 1966 and 1996.[18] These photos open a window to settler entitlement and the mind of apartheid. Almost 50 percent of the 1,446 photographs were highly stylized depictions of white women posing for the camera. Figure 11.1a is representative of this genre of photograph, heralding the arrival of summer and holidays. These images assume a white readership and do the work of normalizing the white beach and entitlement for white audiences. White men were also depicted in bathing gear, but generally in active stances as lifesavers, surfers, and bodybuilders. Black people never featured in representations of Surf City, but they occasionally appeared on the beach as workers, in periphery, as in figure 11.1b. This genre represented the symbolic order of apartheid with beaches being fun-in-the-sun playgrounds for whites along with rugged white masculinity and Black people either entirely absent or captured as lowly workers servicing the economy of leisure. The individuals in the images appearing in

Figure 11.1. Photographs of people on Durban's beaches taken from the *Natal Mercury* between 1966 and 1996. Photo scans by Claudia Plunkett.[19]

the *Natal Mercury* need not have posed self-consciously as white people. They were holidaying, relaxing, suntanning, and so on. So too the persons behind the lens were not necessarily depicting whiteness. Nor were the thronging holidaymakers, surfers, sun tanners, and partygoers on the white playground. Whiteness was the taken-for-granted premise, an invisible racial position, norm, or standard,[20] upon which these forms of subjectivity and personhood were possible. Of course, this racial positioning and subjectivity were predicated on the erasure of Black personhood, autonomy, and ownership of place, which were both taken for granted and enforced.

Disruption: Awakening to Challenge

Perhaps there was a time, at the apex of white supremacy, when racial exclusivity could be taken for granted on the "whites-only" beaches of apartheid South Africa. However, as Cell argues, to function smoothly, the system of white supremacy required "some degree of accommodation and tacit acceptance on the part of those whom it is designed to control."[21] By the 1980s, white South Africa was under siege. Violent and nonviolent struggle against apartheid had gained momentum and the government declared a state of emergency in July 1985 in an increasingly desperate and violent attempt to maintain "peace and order." The summer holidays were a time to escape to the beaches, to free oneself from the stress of politics and the grind of everyday life, but the white playground was under threat by the unwanted intrusion of Black others and the politics of resistance.

During the course of the 1980s, pressure built to accommodate Black holiday makers and tourists. The matter was debated in Parliament and in local government. Protests against "whites-only" signs were met with resistance from residents and holidaymakers who "objected to Blacks on beaches meant for Whites."[22] Durban city created a small section of multiracial beach in 1982, but this did little to quell the protest. Durban resident Morris Fynn repeatedly cut down the apartheid beach signs, for which he was arrested and fined, and the signs were reerected. In 1987, Allan Hendrikse and twenty-five members of his Labour Party took a "protest swim" on a "whites-only" beach in Port Elizabeth.[23] State President P. W. Botha forced him to issue an apology, which provoked outrage "because man does not apologize for what God have [sic] given you."[24] Beaches around the country became sites of struggle under the slogan "All God's beaches for all God's People,"[25] challenging the racial order and unsettling the white playground.

The disruption of "place identity" helped to make whiteness visible as an ideology of place and a form of subjectivity.[26] Transgression of the ra-

cial order of "whites-only" beaches forced a response—"White talk"[27]—
that articulated norms and assumptions of white entitlement that could
previously exist as unspoken norms and assumptions. Kevin Durrheim
and John Dixon presented an analysis of four hundred articles, editori-
als, and letters about Durban's beaches that appeared in the *Mercury*
between 1982 and 1995.[28] The extracts below show two overarching topics of complaint: the sense of being dominated and intimidated by Black
beachgoers and disgust at their conduct.

> We invite up-country visitors to enjoy the festive season on our fair
> (?) beaches [*sic*] and then subject them to the awful experience of be-
> ing dominated and intimidated by hordes of largely undisciplined Black
> people. . . . Looking back a few years, one asks what has happened to the
> happy family groups that one used to see on our beaches.[29]

> I would like to know if it is allowed to get completely undressed and take
> a shower, or lift your dress over your head (no pants on) and wash "you
> know where" in full view of the public? This is the scene at the North
> Beach showers every weekend. If I were to do this I would be arrested.
> If these people want to use the North Beach could they do so with
> decency.[30]

Norms of conduct, dress, behavior, and the correct number and kind
of people on the beach are described in these letters. These are previously unspoken norms against which the behavior of "undisciplined
Black people" or "these people" are judged. The narratives enact forms
of white talk for white hearers just as the images in figure 11.1 depict a
white gaze for white audiences. The "up-county visitors" and the "happy
family groups" are codified refences to white people, as is the anony-
mous but evidently white letter writer, the "I" who would be arrested for
taking an indecent shower.

In all this talk, Black beachgoers are constructed as foreign and out of
place. Figure 11.1c shows how this can be done via visual representation
of norm violation. The Black woman (shot from behind without permission, no doubt) is presented as a spectacle for the white viewer. The contrast between figure 11.1a and 11.1c reveals all that is scandalous: It's not
proper to be dressed in bra and panties. Latch the bra properly! Don't sit

in the water like that. Are you not ashamed to show that body in public? The image of the African woman doesn't fit the genre of sexy, Playboy-Bunny-style white women on the beach. Its purpose is not to herald the summer but to show white readers the new reality of Black people on white beaches. Like the letters considered above, the picture serves as a reminder of an indecent, unwelcome Black presence that undermines the character of the white place. It shouts out "You don't belong here!"

It is a remarkable achievement that European settlers could portray African natives as foreign presences who don't belong on African beaches. This speaks volumes about the nature of white entitlement. It is entitlement to a racial preserve, the beaches as the white playground. This entitlement is premised upon an extensive valuation of whiteness: white is beautiful, white is decent and civilized, white is well-behaved, white is well-dressed, white is normal, white is happy, white is family. White is the standard against which everything is judged. This is what is at stake. White supremacy renders a Black presence in the white playground as a violation of the integrity of the place as well as all that is good and valuable and decent, which are the grounds of entitlement.

These representations of Black presence on "whites-only" beaches not only communicate a mourning for the loss of place, they also expresses outrage at the violation of racial entitlement to the place that is lost. All the narratives and the pictures communicate a self-righteous indignation, umbrage, at the impending upending of the racial order of apartheid. The narratives of disgust underpin anger, even outrage, at the violation of place and the displacement of white people. But this was only the start. There was much more to come.

Displacement

Beach apartheid legislation was repealed in 1989 and the remaining "whites-only" signs were removed. This was a period of massive change in the country as Nelson Mandela was released from prison, the Black liberation parties were unbanned, and multiparty negotiations culminated in the first democratic election in South Africa in April 1994. The newly elected Black majority African National Congress government would ensure public access to public spaces and would criminalize racism. How could white entitlement be expressed in this context?

A number of high-profile cases of racism on the beaches showed the fate of those who would repeat the supremacist views from the past. In January 2016, Durban estate agent Penny Sparrow tweeted her indignation, describing Black beachgoers as "monkeys" let loose on public beaches, "inviting huge dirt and troubles and discomfort to others."[31] Two years later Adam Catzavelos posted a video of himself on holiday on a beach in Greece, reveling in whiteness: "Not one k***r in sight, f**king heaven on earth. . . . You cannot beat this!"[32] Both were convicted of racism and were sentenced to hefty fines of R150,000 each (approximately US$11,000 at the time). Times had changed and the explicit racism of the past was now a criminal offense. White entitlement required new forms of expression.

During the summer holidays of 1999 and 2000, John Dixon and I interviewed white beachgoers at the holiday resort of Scottburgh, sixty kilometers south of Durban. We asked them how their holidays had changed since the repeal of apartheid legislation and to share their experiences of intergroup contact and their opinions about the desegregation of the beaches.[33] The extracts below show the evolution of white entitlement and indignation in the postapartheid context.

> PETER: I'm not a racist . . . It's not about Black or White or something like that. It's like if someone can behave themselves like humans, you know, you'll give them your respect . . . but if someone—White, Black, don't know, it doesn't matter what color— . . . if they behave like animals then I mean I'm against that.[34]
>
> SIMON: I don't mind. If it's safe, I don't mind if we have a toot or two, look after our place. But in a case like yesterday, I mean if you just ride back now, you look at the streets, the streets is full of shit. That, I don't feel comfortable like that, I don't care if Black families is all around us, as long as they keep their space clean, it's fine. If you act like savages then no. Yesterday, where was it? Doonside, you had a guy over there, telling you, he's in the sea, brushing his toothbrush, er, brushing his teeth. I mean that's so stupid.[35]
>
> ANNA: I mean mainly they they've taken over Durban. I don't think that will ever change that's definitely now their little town whatever, beaches, it's theirs. Let them have it, but I think in a couple of years' time Durban is going to be in such a mess. I don't know, that's my opinion. I mean for a holiday I will never ever book any holiday accommodation in Durban itself, forget it, never.[36]

JACK: We do have different ways of doing things you know, let's call it Black and White.

MERLE: Like wearing proper bathing costumes.

JACK: We come here as four people, as a family, whereas bus loads (inaudible) taken over . . . So I'd rather just move away . . . When I was younger, when I was their age we used to holiday in Durban, it was a pleasure, and then it started changing. Eventually we moved further South.[37]

The first two extracts show the continuation of the theme of the indecent presence of Black people on the beach and incredulity at their behavior and appearance. As discussed in the previous section, the spectacle appears as such against the backcloth of assumptions about how beaches "should be" and the norms and assumed entitlements of whiteness. Now these are prefaced with denials of racism—"I'm not a racist," "I don't mind"—and avowals of being colorblind—"It doesn't matter what color [they are]." "White talk" accommodates anti-racism as it articulates entitlement to the (mythical) white playground of colonialism and apartheid.[38]

Mourning of loss and displacement is now accompanied by rhetorical moves to adapt to the new political reality. The second two extracts repeat a well-worn refrain of postapartheid whiteness: "They've taken over. If they want it, they can have it. We will find someplace new." We documented racial patterns of beach occupation for twelve days in 1999 and 2000 and seven days in 2001 and 2002 and observed repeated patterns of racial segregation and "White flight."[39] The experience of loss of place, the sentiment of "they can have it," and the withdrawal and migration this sets up have helped to create the South Africa we see today. New sites of whiteness have been carved out in almost every sphere of life, on the beaches of Ballito and those of Paternoster and in wine farm tourism, edge cities, exclusive clubs, high-end malls, gated communities, equestrian estates, Umhlanga and Sandton, and so on. White people insulated themselves from political change by talk geared to denial of racism and "semigration" to new enclaves of white privilege.

The postapartheid architecture and preserve of whiteness need not be racially exclusive. The white spaces accommodate qualified entry of "decent" Black people who know how to behave. In this way, the norms and standards and other trappings of whiteness are preserved in the present. Apartheid might be gone but whiteness persists in the norms of ("civilized") conduct and ideals of personhood communicated by Californicated, tanned, happy, youthful white bodies along with exclusive holidays and other signifiers of wealth and luxury. A "colonial mentality" of whiteness persists as "residues, appropriations, subconscious reproductions and disguises [which] are diffused in society through subtle and not-so-subtle gestures, attitudes, and informal rules of social relations."[40] These provide a home for whiteness and a place to satisfy and affirm white entitlement in the postcolony. In the process of displacement and semigration, the beach is reconfigured as a retreat for a threatened and threatening deracialized whiteness, an object of contemporary attack and counterattack.

Counterattack

White South Africans have always seen themselves as victims or potential victims. Concerns about the *Swart gevaar* (the Black peril or threat) are condensed in the nightmare of being individually slaughtered in your bed, as depicted by Anton Kannemeyer's print "N is for Nightmare (house)," or collectively being "pushed into the sea."[41] The existential fear of racial violence forms the backcloth to whiteness in settler society.

Elemental anxieties of the nightmare played themselves out on the exclusive Clifton 4th Beach in Cape Town in December 2018. To remove a "bad element" from their wealthy neighborhoods, residents around the beach commissioned a local security company to chase beachgoers off the beach on the afternoon of December 23, 2018. It was a multiracial crowd, but the eviction had all the hallmarks of white entitlement and conveyed reminders of the "forced removals" of apartheid. Black activists responded by conducting a ritual slaughter of a sheep on Clifton Beach on December 29, calling on ancestors to "help us to really cleanse our beaches of racist White people."[42] Andrew Spiegel reports the transcript of an eyewitness account from an unnamed white student:

> The absolute savagery that I . . . saw today. . . . They dragged [the sheep] by the ears into the sea, symbolizing how they are going to drive White people into the sea. . . . They dripped the blood all over the beach . . . to symbolize that "this beach is ours" and they will do anything to claim

this land which they call their land, even though their people are not indigenous in this part of the world. They . . . sprayed [the sheep's] blood all over the beaches to mark that this is now their territory and they are going to drive White people from Clifton into the sea and they're going to come and expropriate and take all the land, all the houses . . . here.[43]

All the elements of the nightmare are there: the savagery, the violence, the blood and sacrifice, and the driving of white people into the sea. The nightmare is itself permeated with entitlements of whiteness, which it expresses as much as consoles. The old themes of desecration of white space, incredulity, savagery, and the sense of self-righteous indignation attend these narratives of victimhood.

Another recent incident, also in Cape Town, shows how white victimhood is evolving. We pick up the story of whiteness on the beaches in the context of the COVID-19 pandemic. Emergency Level 5 Lockdown rules saw all beaches in the country closed from March 26 to 1 May 1, 2020; beach gatherings were banned until May 31, 2020; and a nighttime curfew on beaches was imposed until August 18, 2020. After the first wave of the pandemic, beaches were opened and then closed again during the second and third waves, especially over the peak holiday season in December 2020 through January 2021. There was pushback against the government's lockdown restrictions throughout the pandemic, especially the ban on alcohol and tobacco sales, and this spilled onto beaches in January 2021. Activist groups going by names such as Woke Nation and We Are More (WAM) organized protest gatherings on the beaches in violation of the lockdown (figure 11.2). The rationale behind the protests was explained by an organizer, Clay Wilson: "We want beaches opened because they belong to the people. People are starving. They can't earn a living because of COVID-19. The government must open beaches and drop the liquor ban. These COVID-19 laws are only serving a few, who are politically connected, and affect those who earn a living through tourism."[44] This is a new incarnation of white talk. Under apartheid and into the postapartheid context, white people have often stood against "the people," objected to the presence of Black people on "their" beaches, and withdrawn into new white enclaves rather than integrate. Multiracial beaches, which are now acknowledged to belong to "the people" become places to speak on behalf of "the people" against the govern-

Figure 11.2. We Are More protest at
Fish Hoek beach, January 2021.
Photo courtesy of a Facebook user.[45]

ment. This is "white talk" reiterating familiar themes of "decay, corruption, greed and incompetence" in government.[46] Freshly rehabilitated by their perceived victimhood, the supporters and defenders of apartheid now position themselves in the vanguard of the struggle for justice. The deracialized language is careful to communicate the entitlements of victims, repeating the self-righteous indignation of the master's voice.

The privilege and entitlements of whiteness were apparent to Black commentators, including provincial chair of the political party the Economic Freedom Fighters, Melikhaya Xego, who contrasted the policing of the beach gathering and a gathering of Black "pensioners and people with special needs" a week earlier: "Based on the previous experiences, it is clear to us that if the majority of these protesters were Black they would have been arrested and shot at with stun grenades and water cannons by the police and law enforcement, but because the majority are White the police did nothing."[47] The story ends with pictures of a WAM leader, Craig Peiser, being arrested and the report of his disruptive court appearance and his subsequent admission to Valkenburg hospital to where he was declared unfit to stand trial, being unable to "distinguish between right and wrong."[48] As with the well-publicized arrests of and fines imposed on Penny Sparrow and Adam Catzavelos, this is a cautionary tale that sets limits for white entitlement.

An Adaptable, Intersectional Palette of Whiteness

The story of settler whiteness has been recounted here as a geographically located historical narrative. Both the geography and history require qualification. Fixing attention on the beach has helped to contain the narrative, to tell a big history in a few pages. Although beaches might appear peripheral to the big questions of land and urban justice, they teach us much about the morphing persistence of white supremacy. The framing of civilization versus savagery, the desire for exclusion, and the self-righteous indignation that define white entitlement across all four historical contexts considered here also find expression in other spheres of life. Whiteness is enormously adaptable. No doubt, similar kinds of "white talk" can be found off South African shores in postcolonial contexts that are now controlled by European settlers (e.g., in the United States, Australia, Europe) and those which are not (e.g., in Africa and Asia).

The historical narrative gives the impression that the shades of whiteness are stages of whiteness, beginning at the highest stage of white supremacy and then following its fall. However, this is not a developmental account of whiteness. All four shades of whiteness can occur together in any particular context. They blend together as a palette of whiteness with all expressions occurring at once at this point in history. Segregation ensures that white playgrounds continue to proliferate in postapartheid South Africa. Qualified Black people can have entry and so there is always the threat of disruption, the wrong type, violence, and displacement and therefore the need to fight back to make South Africa great (or white) again.

The focus on entitlement helps us appreciate the adaptability and intersectionality of whiteness. The metaphor of the "invisible knapsack" reminds us that whiteness owns privilege.[49] Semiotic privilege links desirability and value to whiteness, making whiteness an aspirational standard. This privilege can also be cashed in for favorable treatment—for example, by policing the white crowd with kid gloves. However, this is not always the case. White talk loves the consolation of victimhood precisely because this points to sites and instances of loss of privilege and occasions where white people are special targets of attack. At these "intersections of whiteness,"[50] where being white can be a liability as much as an asset, the entitlements of whiteness may trump its privileges in mapping the way forward. The entitlements of whiteness are an adaptable, intersectional palette.

NOTES

1 Ana Deumert, "Beachspaces: Racism and Settler-Colonial (Im)Mobilities at the Shoreline," in *Exploring (Im)mobilities: Language Practices, Discourses and Imaginaries*, ed. Anna De Fina and Gerardo Mazzaferro (Bristol: Multilingual Matters, 2021), 183–206.

2   Peggy McIntosh, "White Privilege: Unpacking the Invisible Knapsack," in *Revisioning Family Therapy: Race, Culture, and Gender in Clinical Practice*, ed. Monica McGoldrick (New York: Guilford, 1989), 147–52.

3   Paul Gilroy "Joined-up Politics and Postcolonial Melancholia," *Theory Culture Society* 18, no. 2–3 (2001): 151–67, https://doi.org/10.1177/02632760122051832.

4   Evangelia Kindinger and Mark Schmitt, *The Intersections of Whiteness* (Abingdon, UK: Routledge, 2019).

5   Gillian Straker, "Unsettling Whiteness," in *Race, Memory and the Apartheid Archive*, eds. Garth Stevens, Norman Duncan, and Derek Hook (Johannesburg: Wits University Press, 2013).

6   Sibusiso Maseko and Kevin Durrheim, "Simmering Hostilities, Group Identity and Contested Autochthony Beliefs in Settler Societies," in *The Psychology of Politically Unstable Societies*, eds. Barbara Lášticová and Anna Kende (Abingdon, UK: Routledge, 2023).

7   John Whitson Cell, *The Highest Stage of White Supremacy: The Origins of Segregation in South Africa and the American South* (Cambridge: Cambridge University Press, 1982).

8   Anthony Lemon, "The Apartheid City," in *Homes Apart: South Africa's Segregated Cities*, ed. Anthony Lemon (Indianapolis: Indiana University Press, 1991), 2–26.

9   Murriel Horrell, *A Survey of Race Relation in South Africa 1954–1955* (Johannesburg: South African Institute of Race Relations, 1955).

10 Orvar Löfgren, *On Holiday: A History of Vacationing* (Berkeley: University of California Press, 1999).

11 Franco Ferrario, "Black and White Holidays: The Future of the Local Tourist Industry in South Africa," *Annals of Tourism Research* 13 (1986): 331–48.

12 Christian Rogerson and Zoleka Lisa, "'Sho't Left': Changing Domestic Tourism in South Africa," *Urban Forum* 16, no. 2–3 (2005): 88–111.

13 Valerie Møller and Lawrence Schlemmer, *Attitudes Toward Beach Integration: A Comparative Study of Black and White Reactions to Multiracial Beaches in Durban.* (Durban, South Africa: Centre for Applied Social Sciences, University of Natal, 1982).

14 Robert Preston-White, "Constructed Leisure Space: The Seaside at Durban," *Annals of Tourism Research* 28, no. 3 (2001): 581–96.

15 Glen Thompson, "Reimagining Surf City: Surfing and the Making of the Post-Apartheid Beach in South Africa," *International Journal of the History of Sport* 28, no. 15 (2011): 2115–29, https://doi.org/10.1080/09523367.2011.622111.

16 John J. Bukowczyk, "California Dreamin', Whiteness, and the American Dream," *Journal of American Ethnic History* 35, no. 2 (2016): 91–106.

17 Thompson, "Reimagining Surf City.

18 Claudia Plunkett, "Regulating the Body: Images of Men and Women on the Beach" (Master's thesis, University of Natal, 1998).

19 These images are newspaper clippings, historical documents that demonstrate the crude racist and sexist habitus and representation of people on South African beaches under apartheid.

20 Richard Dyer, "The Matter of Whiteness," in *White Privilege: Essential Readings on the Other Side of Racism*, 2nd ed., ed. Paula Rottenberg (New York: Worth Publishers, 2005), 9–14.

21 Cell, *Highest Stage of White Supremacy.*

22 Carole Cooper, Shireen Motala, Coleen McCaul, Thabiso Ratsomo, and Jennifer Shindler, *Survey of Race Relations in South Africa 1983* (Johannesburg: South African Institute of Race Relations, 1984).

23 The Labour Party represented the community members of color in the tricameral Parliament of the time.

24 Labour Party MP Peter Mopp, cited in Carole Cooper, Jennifer Shindler, Colleen McCaul, Robin Hamilton, Mary Beale, Alison Clemans, Lou-Marié Kruger et al., *Survey of Race Relations in South Africa 1986* (Johannesburg: South African Institute of Race Relations, 1987), 738.

25 Dene Smuts and Shauna Westcott, *The Purple Shall Govern: A South African A to Z of Nonviolent Action* (Cape Town: Oxford University Press, 1991).

26 John Dixon and Kevin Durrheim, "Displacing Place-Identity: A Discursive Approach to Locating Self and Other," *British Journal of Social Psychology* 39 (2000): 27–44.

27 Melissa Steyn and Don Foster, "Repertoires for Talking White: Resistant Whiteness in Post-apartheid South Africa," *Ethnic and Racial Studies* 31, no. 1 (2008): 25–51.

28 Kevin Durrheim and John Dixon, "The Role of Place and Metaphor in Racial Exclusion: South Africa's Beaches as Sites of Shifting Racialization," *Ethnic and Racial Studies* 24 (2001): 433–450.

29 Letter to the editor, *Mercury*, January 13, 1988, 10.

30 Letter to the editor, *Mercury*, November 9, 1990, 9.

31 City Press, "Racism: Penny Sparrow Fined R150K, Community Service for Theunissen," *News24*, June 16, 2016, www.news24.com.

32 Ntwaagae Seleka and Riaan Grobler, "Adam Catzavelos gets Suspended Sentence for K-Word Rant," *News24*, February 28, 2020, www.news24.com.

33 Kevin Durrheim and John Dixon, *Racial Encounter: The Social Psychology of Contact and Desegregation* (Oxford: Routledge, 2005).

34 Durrheim and Dixon, *Racial Encounter.*

35 Durrheim and Dixon, *Racial Encounter.*

36 Kevin Durrheim, "Socio-Spatial Practice and Racial Representations in a Changing South Africa," *South African Journal of Psychology* 35, no. 3 (2005): 444–59.

37 Durrheim and Dixon, *Racial Encounter.*

38 Melissa Steyn and Don Foster, "Repertoires for Talking White: Resistant Whiteness in Post-apartheid South Africa," *Ethnic and Racial Studies* 31, no. 1 (2008): 25–51.

39 Durrheim and Dixon, *Racial Encounter.*

40 Moses E. Ochonu, "Looking for Race: Pigmented Pasts and Colonial Mentality in 'Non Racial' Africa," in *Relating Worlds of Racism: Dehumanization, Belonging and*

the Normativity of European Whiteness, ed. Philomena Essed, Karen Farquharson, Kathryn Pillay, and Elisa J. White (Cham, Switzerland: Palgrave Macmillan, 2019), 3–38.

41 Cornel Verwey and Michael Quayle, "Whiteness, Racism, and Afrikaner Identity in Post-Apartheid South Africa," *African Affairs* 111, no. 445 (2012): 551–75.

42 Andrew D. Spiegel, "Sheep, Herbs and Blood on the Beach: Discrepant Representations of Ritual Acts for Essentialising and Reinforcing Difference in Contemporary South Africa," *Anthropology Southern Africa* 43, no. 2 (2020): 143–55, https://doi.org/10.1080/23323256.2020.1740605.

43 Spiegel, "Sheep, Herbs and Blood on the Beach."

44 Nicole McCain and Ntwagae Seleka, "COVID-19 Regulations Cape Protestors Say Enough Is Enough, Want Beaches Reopened," *News24*, January 30, 2021, www.news24.com.

45 Shakirah Thebus, "Group Involved in Beach Protest Calls on Public to Join End Lockdown Demonstration," *IOL*, February 5, 2021, www.iol.co.za.

46 Kim Wale and Don Foster, "Investing in Discourses of Poverty and Development: How White Wealthy South Africans Mobilise Meaning to Maintain Privilege," *South African Review of Sociology* 38, no. 1 (2007): 45–69, https://doi.org/10.1080/21528586.2007.10419166.

47 Okuhle Hlati, Okuhle, "If the Majority of Protestors Were Black, They Would Have Been Arrested and Shot At" *IOL*, February 1, 2021, www.iol.co.za.

48 Staff Writer, "Man Accused of Assaulting ENCA Journalist Declared Unfit to Stand Trial," THEWC.co.za, February 20, 2021, https://thewc.co.za/man-accused-of-assaulting-enca-journalist-declared-unfit-to-stand-trial/.

49 Peggy McIntosh, "White Privilege: Unpacking the Invisible Knapsack," in *Revisioning Family Therapy: Race, Culture, and Gender in Clinical Practice*, ed. Monica McGoldrick (New York: Guilford, 1989), 147–52.

50 Evangelia Kindinger and Mark Schmitt, *The Intersections of Whiteness* (Abingdon, UK: Routledge, 2019).

12

"A Coveted Paradise"

Policing and the Devastation of Brazil's Coastal Lands

KEISHA-KHAN Y. PERRY

US Black feminist scholar bell hooks (1952–2021) gave us innumerable conceptual tools to understand the complexities of race, gender, and place. Her critical essay "Homeplace (A Site of Resistance)," first published in 1990 in her *Yearning: Race, Gender, and Cultural Politics*, illustrates how the homeplace can represent spaces of oppression as well as liberation.[1] This chapter provides a personal reflection on the interconnectedness of policing and evictions in the northeastern Brazilian city of Salvador, Bahia. With careful attention to the threat to Black dignity and humanity, I explore how neighborhoods where the cultural imagination and radical politics flourish even as poor Black people experience the brutality of white supremacist violence and spatial displacement. In hooks's formulation, oppressed people make their homes in inhospitable places, resist gendered and racial domination of space, and demand to have a sense of cultural and political belonging. In Bahia, beach lands, including in the historic city center, were historically undesirable wastelands surrounding plantations and military forts on which Indigenous, formerly enslaved, and free people have built communities. I describe the context of how social movement activists fight to keep as Black home spaces beach lands where they have forged communities and survived amid violence for generations.

Gamboa de Baixo, Brazil

After a blazing hot day with the sun beating down on us on the Preguiça beach, we're ready to go home. The old residents of Preguiça have long been removed from the land near the beach, moved to a neighborhood about two hours away during the construction of the sculpture park at the Museum of Modern Art. By the late afternoon, local free diver and fisherwoman Denise convinces us that it's better to take the boat back to Gamboa de Baixo rather than walk home on the winding road along the Bay of All Saints. We climb into the rowboat close to shore without getting stuck in the sand. The children play soccer on the beach, the fishermen cast their nets, and the retaining walls around the yacht club are silhouetted against the orange sky—everyday life on the Bay of All Saints. Poor and rich living and playing side by side. Preguiça was one beach that was still available to Denise and her neighbors. No public transportation is needed to get there and the sun doesn't burn your skin on the boat as much as walking on the hot asphalt; at least, that's what I've come to believe. Saltwater and seaweed made taking deep breaths easier, unlike the smell of human excrement hidden behind the concrete walls that separated us from Contorno Street, which runs over the Gamboa de Baixo neighborhood in the center of Salvador. Families at the yacht club play in the pool and board their boats, and we wave to a marinheiro, a sailor, married to one of the Gamboa activists, who is steering a yacht. "You ate well today, eh?" we call to him. Denise is straining at the oars when we see her uncle coming toward us in a boat. We gladly accept the tow he offers. We'll get home in no time. But the yacht is going so fast that water is splashing into our boat. Denise waves to her uncle to slow down, but he doesn't see her. "Jump overboard!" she yells at me, afraid that the boat is about to capsize. I can't hear her over the sound of the waves and the motorboat. "What?" I shout back at her. "You're going to have to jump!" she says. "Jump where?" I ask. "In the water!" she answers. "But I can't swim!" I say matter-of-factly. "*Ta brincando*! You're joking!" she exclaims, incredulous. Everyone knows how to swim in this community built on the bay. Denise bursts out laughing. She has known me for all the years I've been doing research in this coastal community but never knew I couldn't swim. She thinks I'm lying but she doesn't want to take a chance that I'm not. Finally, she catches her uncle's attention and tells him to slow down and take us to the boat ramp in Gamboa. "How is it that you don't know how to swim?" she asks. I am indeed from a small island in the Caribbean. I've been on these fishing boats almost as long as I've been on land. But I had never learned to swim. I feel ashamed revealing this to Denise. At the boat ramp, I climb out of the boat and vow that by the time I return to Gamboa the following summer, I will be a swimmer, like the native Gamboans I have come to know and love while studying their ongoing fight to keep possession of their ancestral coastal lands. ·

The Everyday Reality of Rogue Waves

My interest in conflicts over beach lands begins in an unlikely place. Not in Jamaica, where I was born and spent my early years. In fact, in Jamaica, I mostly went to local rivers and swimming holes. I could catch shrimp with my bare hands, but I never learned to swim. Beaches were far from the countryside and reserved for special occasions like the school trips to Dunn's River Falls. They were for the white foreigners who we hardly saw in the communities with names like Mt. Friend- ship, Golden Spring, and Stony Hill, where shared taxis and motorcycles were the only alternatives to walking. I have only vague memories of the beaches in Ocho Rios, where my mother worked at a luxury hotel before it was washed away by Hurricane Alan in 1980, the same hurricane that destroyed the famous Trident Hotel in Port Antonio, my father's home- town. On school trips, I had also visited Port Royal in Kingston, called "the wickedest city on earth" for its history of pirates and naval battles and its leaning buildings, the aftermath of the 1692 earthquake and giant tidal wave that consumed most of the city, which was further devastated by later hurricanes and fires. As a child, the beaches of Jamaica felt both marvelous and yet a part of everyday life.

I began to think more seriously about beaches when my mother signed me up for a middle-school creative writing summer camp that took me to Sandy Hook beach in Highlands, New Jersey, for inspira- tion. This was not a beach, I decided. Through prose, I was determined to illustrate the dark color of the water (did people really know what strange things lurked at the bottom?), the cold temperature (it was not meant for humans), and the medical waste (needles were knowingly being discarded from New York City hospitals into Raritan Bay in the southern section of Lower New York Bay). My early writing on beaches as a middle schooler described vast wastelands that contrasted starkly with the paradise I had imagined possible in the United States. I could not narrate stories of beaches fit to be enjoyed, and my environmental activism began on those New Jersey shores.

It was in Brazil in the late 1990s that I first saw people, Black people specifically, going en masse to the beach for leisure, packing city buses and leaving sand wherever they stepped. But I did not enjoy the first beach I visited in Rio de Janeiro when I was in college. It was not the beach where the *povo* (people) carried their coolers and precooked meals. The organizers of my study-abroad program took us to a private beach in a gated residential community on the outskirts of Rio. The housekeepers and gatemen were the only Black people I saw. I sat uncomfortably with my feet in the sand and watched the beautiful blue water framed by mountains in the distance: Sugarloaf Mountain, Pedra Bonita, Pedra da Gávea, and Tijuca Peak. No one on this beach was selling *caipirinha* cocktails, *queijo coalho* (roasted cheese), foot massages, body oil, or bikinis, things sold on most public beaches throughout Brazil. That day, the waves came and went, touching my feet on occasion, warmer than the waters of Rio de la Plata in Argentina, where I had been living during the school year, and the sand was finer than at Sandy Hook. Then, in a matter of seconds, something apparently unusual happened in what felt like slow motion, incredible to fathom, as I sat and watched. My classmates chatted, unaware of what was about to happen. A large rogue wave suddenly washed away the shoreline and buried our belongings in the sand. It felt like I was also moving in slow motion as I alerted my classmates. We moved as fast as we could to get up to the rocks, save ourselves, our clothes, and the bag holding our wallets and purses.

That moment on that private beach, when I was aware of a grave danger and my white classmates were not, summed up my first visit to Brazil and the many years of my racial reality and experience with the built environment that followed. The *vai e vem* (coming and going), the apparent cordiality, the gates, the walls, the service and social elevators side by side, the hypervisible Black housekeepers and the doormen, and the unremarkable events of everyday life and structural racism all coexist. The fact is that this perverse racial reality in Brazil is unpredictable but can be very dangerous for the Black people in its path who see it clearly as it gains momentum. The police raids that coastal communities like Gamboa experience, while a part of everyday life, are always sudden, violent, and extremely unpredictable in terms of the devastation and the "slow death caused by sequelae" that Christen Smith describes as the "particular weight of anti-Black state violence."[2] Actual rogue waves exist in places like Bahia, but they are understood as manmade destruction caused by the building of the artificial barrier reefs for the yacht clubs or the removal of sand to make space for massive cruise and cargo ships. Amid any portrait of a paradise in Brazil, like the unexpected wave in Rio de Janeiro, I am always awakened, made conscious, by a different racial reality.

Angela Gilliam's Debunking of the Tropical Utopia

In Angela Gilliam's classic essay "From Roxbury to Rio and Back in a Hurry" (later expanded as "Black and White in Latin America" and republished in David Hellwig's 1992 volume *African-American Perspectives on Brazil's Racial Paradise*[3]), she vehemently rejects the image of Brazil as a racial democracy and challenges the notion that a paradise exists there for Black people. This view was a correction to previous generations of Brazilian and North American scholars (E. Franklin Frazier and Cyril Briggs, for example) who espoused the idea that a tropical utopia existed for people of color in Brazil, creating new possibilities for African Americans, such as W. E. B. Du Bois leading collective efforts to flee racial terror in the United States and settle in Brazil. "Brazil: Do You Want Liberty and Wealth in a Land of Plenty? Unlimited Opportunity and Equality? Then Buy Land in South America," read the advertisement by the Brazilian American Colonization Syndicate (BACS) of Chicago published in Du Bois's newspaper, *The Crisis*, in March 1921.[4] It is probable that African American visitors to Brazil at the turn of the twentieth century misunderstood the lack of legalized racial segregation and their ability to travel with ease as lighter-skinned, educated Black male elites as representative of an absence of racial strife or of the overall experiences of the majority-Black population. In 1950, when African American dancer and anthropologist Katherine Dunham was refused a room at the Hotel Esplanada in São Paulo, she was not mistaken to identify it as a clear case of racial discrimination. The case generated much public attention and intense debate in the press and shaped the passing of the Afonso Arinos Law barring racial discrimination in the public sphere and public accommodations.[5] As Brazilian social scientists (Cidinha da Silva, Marielle Franco, Lélia Gonzalez, and Sales Augusto dos Santos, to name a few[6]) would document in the years that followed, anti-Black racism was a pervasive problem in the country.

232 KEISHA-KHAN Y. PERRY

Gilliam argued for a hemispheric approach to understanding anti-Black racial oppression that paved the way for my essay in progress on the 2020 Paraisópolis police massacre, which demonstrated the long duration of violence rather than paradise in the metropolis of São Paulo. Gilliam offered a more expansive critique of a racial paradise at the core of Christen Smith's 2016 book, *Afro-Paradise: Blackness, Violence, and Performance in Brazil*. As Smith states, while Bahia is described as the land of happiness, it is "one of the most violent places to live in Brazil, particularly for Black youth." She continues, "The realities of state violence require that we rethink Bahia's image as an exotic black space."[7] I draw from this analysis of the "Afro-paradise" to explore how in majority-Black Brazilian cities like Salvador, popularly known as the Black Mecca in the African diaspora, social, political, and economic spaces remain segregated. While public and still frequented by the local population, beaches have become increasingly privatized by yacht clubs, hotels, and luxury apartment piers. Black people belong in these spaces as laborers (security guards, housekeepers) and cultural workers (*baianas de acarajés* [street vendors who sell bean pies fried in palm oil], *capoeiristas*, musicians, etc.). Denise, my boating companion, has been a diver for a local hotel who cleans the spiny sea urchins off the ocean floor.

In recent decades, more US Black scholars of Brazil, such as Kia Lilly Caldwell and Okezi Otovo,[8] have been become preoccupied with confronting the disparities that resemble gendered, racial, and class-based patterns in this country. Over the past century, African American writings on Brazil have shifted from positive affirmations of a racial paradise to critical perspectives on the generalized negation of Blackness, treatment of Black people as nonhumans, and a lack of democratization of citizenship rights and resources in Brazil.[9] This scholarship affirms an analytical shift toward a focus on the differences and specificities in socioeconomic realities structured by race and gender, such as the militarization of urban policing and mass land evictions. Here, I contribute to that ongoing exploration of the similarities between, and the globalization of, mechanisms of gendered racial subjection and injustices that challenge national narratives of social progress and racial democracy. The transnational emphasis on land dispossession and Black women's leadership and participation in the ongoing struggle for land rights shows us the devastating impact of anti-Black systems of domination as well as the collective political work to transform these societies. Looking at beach lands specifically, Andrew W. Kahrl writes in *The Land Was Ours: How Black Beaches Became White Wealth in the Coastal South* that "the history of African American coastal land ownership begins in the decades following emancipation,"[10] and similarly, in the case of Brazil, "coastal regions that once constituted the heart of slave power in the United States witnessed a profound transformation."[11]

In Brazil, forced removals of the local Black population to the distant periphery and the reorganization of public transportation makes the sojourn to the city's beaches more difficult. Katia Costa Santos, in her essay "Black Girls in Ipanema," written for a special issue on housing that I organized for the *NACLA Report on the Americas*, describes how difficult it was for her cousins to get to the beach for the first time: "Their town has several beautiful waterfalls with which they are quite familiar, but no beaches. Coming to Ipanema, for example, from Duque de Caxias is an ordeal, like it was for them that day: it takes at least two buses, or two buses and a subway, or two buses and a train."[12] She also writes that "segregation is even greater in Rio today, due to what the authorities call 'city development.'"[13] Hence, with the rapid development of the city with more modernized public transportation such as the metro, Black people are forced to live farther and farther away from the coast. The photographs in Santos's stunning essay of her cousins standing in awe at the vastness of the sandy terrain of the beach and blue water illustrate the paradox of Rio as a *cidade maravilhosa*, a marvelous city. Bloody clashes between people and bulldozers, residents being forced to relocate, and the police leveling communities defined the years leading up the 2014 World Cup and the 2016 Olympic Games. The community of Vila Autódromo, home to over seven hundred families and located along the Jacarepaguá Lagoon, was forced to move to the distant periphery. In a 2020 article in the *Washington Post*, a resident stated, "The lagoon is still there, but residents say that since the construction for the Olympics began, it's been too polluted to fish there. The view from the western shore, meanwhile, is desolate."[14] Beneath the surface of the beauty of these coastal cities in Brazil lies the violence of racial segregation, forced removals, state-sanctioned police terrors, and the long duration of negation of "Black joy and the power of claiming the right to the city" all across Brazil.[15]

In Defense of a Black Paradise

In Salvador, Bahia, before the mid-1990s, when government officials started marking their houses for removal in order to make way for development and gentrification, residents of the coastal lands of Gamboa de Baixo generally say that they did not know that they were living in any sort of "paradise." While children draped seaweed on their heads and pretended to be mermaids, the neighborhood located on the Bay of All Saints lacked basic infrastructure such as indoor plumbing and sewers. The residents were living in social conditions that were well behind the development of the rest of the city. Gamboa is located next to another seaside community, Solar do Unhão, built on the lands of a former sugar refinery. Coastal and beach lands were considered undesirable lands, reserved for the Black laborers from the plantations and refineries and for those who worked in the mansions of the upper city. A section of the Gamboa neighborhood was built inside the ruins of the São Paulo da Gamboa Fort, built in 1646. An old canon remains intact inside and it was used to salute the arrival of the Portuguese royal family in 1808. For the last three decades, urban planners and the Bahian department of tourism have been trying to develop the land for maritime heritage tourism, which would entail removing the families that have been living in the fort for over a century with the permission of the Bahian Navy after they abandoned it for military purposes. For coastal residents, slavery and the colonial past did not seem like distant memories. Completely surrounded by yacht clubs and luxury residences and hotels, Gamboa residents have resisted the removal of any families as they fear tourism in the fort will lead to the removal of the coastal neighborhood surrounding the fort.

In the 1990s, the seaside community of Preguiça, located on the beach next to Solar do Unhão, was forcibly moved to the distant periphery of Salvador. The government built a sculpture park, now located next to a new modern art museum and a beachfront restaurant serving Bahian cuisine, with waves and sunsets in the background. Residents witnessed the expansion of the Bahia Marina yacht club and the construction of more luxury apartments overlooking the bay. They gained access to urban development plans to organize a maritime tour of the bay that would include a stop on Gamboa lands, specifically inside the São Paulo da Gamboa Fort. The plans include historical restoration for tourism as well as restaurants, just like the art museum that was a former sugar refinery. The construction of piers that resemble the nearby yacht clubs and apartment buildings would facilitate access from the water. With the circulation of these plans, residents began to take seriously the fact that they were living in what they started to call *um paraíso cobiçado*, a coveted paradise. The beach they played on after school and work, the waters that provided food for their families and spiritual rejuvenation, were destined for tourism and luxury real estate development, not for poor Black people to live next to and enjoy. Undesirable coastal lands of colonial Brazil where poor Black people settled, built their homes, and cultivated sustainable relationships with the sea were now "coveted" lands required for desired luxury and leisure tourism around the bay. Like numerous other neighborhoods that have since disappeared along the Bay of All Saints, Gamboa de Baixo, located on the socio-spatial margins of the city, on the coastal lands, has become part of the coveted center of tourism and residential development.

I am particularly interested in how Gamboa activists facing the threat of displacement forged a struggle that constructed an image of living in paradise, an image distinct from the state's portrait through the media of rampant drug trafficking, prostitution, and indiscriminate violence. The land was desirable, not the people; urban developers oftentimes repeated in meetings, without apology, "Gamboa is the face of Bahia, but it's not a place for poor Black people to live." In recent years, I have noted a more significant politicization of a language of paradise, turning the state's logic on its head. This could be understood as a necessary response to the reality that Smith describes as "a routine politics of gendered racialized terror toward the majority-black working class that manifests in the systemic killing of black people by the police throughout the country."[16] Local activists have described the militarized police terror that is part of the process of displacement as a result of assassinations, forced banishment (forcing people to flee the violence), and incarceration. In a city with the largest concentration of African descendants, second only to Lagos, Nigeria, and oftentimes referred to as a "Black Mecca"[17] or "Black Rome,"[18] Black people have become unwanted in the coveted paradise and face violent removal. As I highlighted in the introductory essay of the *NACLA Report on the Americas* special issue on housing justice in the Americas,[19] Tuesday, March 1, 2022, was a brutal reminder of this kind of racialized terror endemic in poor Black neighborhoods. During an early morning invasion in Gamboa de Baixo, the Bahian state military police executed three young people: sixteen-year-old Patrick Sapucaia, twenty-year-old Alexandre Santos dos Reis, and twenty-two-year-old Cleberson Guimarães. Tuesday afternoons were when I usually met with three Brazilian colleagues—longtime Gamboa de Baixo activist Ana Cristina da Silva Caminha, urbanism graduate student Matheus Tanajura, and literature professor and writer Claudia Santos—to discuss our graphic novel project titled *A Coveted Paradise*. The project illustrates the lives and political stories of the mostly women activists and collaborative scholars who have dedicated their lives to Gamboa de Baixo's struggle. For three decades, these women have battled developers and the police to resist forced displacement and to claim their rightful place on the city's coastal lands. The deaths of Patrick, Alexandre, and Cleberson were a heartbreaking reminder for Gamboa residents and poor Black communities in general that the violent siege on their "coveted paradise" is ongoing. But so, too, is their resistance. Resistance and fear exist side by side, and I often wonder how Black women find the courage to speak out against the terror of everyday life. "Somos gente," they say, we are people. The poor Black women who courageously reject the police version of events in neighborhoods like Gamboa de Baixo highlight the political urgency of collectively addressing police violence alongside issues of public health, education, immigration, housing, land, and urban infrastructure.

In working with Gamboa activists for over two decades, I have learned to take a more nuanced approach to analyzing the grassroots development of the discourse of the *paraíso cobiçado* and the political possibility of the preservation of autonomous spaces and, in the process, Black lives. Gamboa residents now understand the term *paraíso cobiçado* as a defining linguistic turn in the social movement. In our research for the graphic novel, we have interviewed activists who recount stories of an ideal coastal paradise with long coastlines, thriving seaweed to protect the sea life, predictable tides, and abundance of fish (since deteriorated after the expansion of the yacht club). They now want the paradise for themselves—but without the past racial-class logics of abandonment, socio-spatial isolation, and threat of displacement to the distant periphery. The construction of the coveted paradise is the key way that they have made claims to the city and the coast, being critical of development logics of demolition while demanding access to modern infrastructure and redefining Black personhood. In this vein, the idea of the coveted paradise also challenges state violence in the form of militarized policing that seeks to clear both the landscape and the people who have inhabited the lands. In 2021 and 2022, Bahia, with an 80 percent Black population, was the Brazilian state where the most people were killed by the police. The majority of the victims were Black and from neighborhoods like Gamboa de Baixo.

I continue to write about the Gamboa de Baixo struggle in solidarity with social movement activists who keep these beach lands as Black spaces, where they have forged communities and survived amid the violence for generations. Politicizing the coastlands of Salvador as a paradise to be coveted as places for poor Black people reveals how Black people have subverted the devastation of the barracoons, the plantations around the bay, and present-day genocidal policing. I am interested in advancing the idea of how this claim to paradise can be deemed radical, similar to what Kevin Quashie calls a move to exhibit claims to aliveness and Katia Costa Santos describes as an urgent demand for territorialized joy.[20] Claiming paradise is radical in a country where death is systemic and deemed inevitable and in a city where Gamboa is one of just a few coastal communities that remain after decades of bulldozing, removals, and violent policing. Paradise, in essence, has become remarkable in the realm of grassroots politics. It is where famous artists now want to film music videos and frustrated city residents could find refuge in a pandemic. Paradise becomes a way of claiming the historical right to the city and building a modern city that includes poor Black people rather than eliminating them.

The Face of Bahia

I began this essay with bell hooks, and I want to end this discussion of the ongoing struggle for beach lands in Bahia with a phrase from her canonical text "Homeplace: A Site of Resistance." The essay offers an ode to Black women that best sums up the political spirit of my scholarly and activist work and the political spirit of this chapter: "I want us to honor them, not because they suffer but because they continue to struggle in the midst of suffering, because they continue to resist."[21] As a result of persistent struggle, Gamboa de Baixo remains one of the few Black communities occupying the coastal lands of the Bay of All Saints, where luxury apartment buildings and yacht clubs have replaced traditional fishing villages. The ongoing state-sanctioned killings of Gamboa residents and the willingness of community members to continue to stand up against the violent siege of their "coveted paradise" is the key reason that these communities continue to thrive despite the persistent destruction of the built environment, the sea, local customs, and human lives. This is an example of what happens when Black people resist being treated as unwanted guests on the land and, as Isabel Wilkerson writes, "find their proper place in the sun."[22] As hooks affirms, it is important to honor the poor Black women workers with little formal education who live and die in these coastal neighborhoods and who are leading conversations about the genocidal nature of the simultaneous disappearance of Black people and Black spaces in cities. Courtney Morris's book *To Defend This Sunrise: Black Women's Activism and the Authoritarian Turn in Nicaragua* similarly emphasizes the myriad ways that Black people are dreaming of liberation and autonomy, oftentimes tied to territorial belonging.[23] As Geri Augusto writes in her essay in homage to Marielle Franco, "In the Brazilian Northeast, along the sea-curving Recôncavo and in the many rivers of the interior, are fishing quilombos (maroon communities) where women and men are waging a struggle for human rights and territory—for land, waterways and the ways of life and knowledge that shape them as a people—a struggle that is as difficult and important as that of the favelas."[24]

As a feminist anthropologist documenting routinized violence of displacement in coastal communities in Brazilian cities, I end with a provocation: What would our focus on social justice in the city look like if we gendered and racialized the conversation? Black activists who have only known these now coveted beach lands as their homes are pushing the boundaries of these political debates about the violence involved in planning for the Olympic Games and the World Cup in coastal cities like Rio and Salvador, Brazil. A significant part of understanding the gendered dimensions of beach politics transnationally is to look at how Black people, poor people, and women are fighting for rights and resources such as access to the coast and the sea. Specifically, they defend the human right to housing and to living in their coastal neighborhoods with dignity and without violence in cities. In the case of Salvador, they have weaponized the discourse of "paradise" to assert that "Gamboa is the face of Bahia, and it is the right place for Black and poor people to live." What I have seen emerge in Gamboa de Baixo, from the collective laughter on the fishing boats to nicknames such as "Sereia" (Mermaid) to the bikini as a routine form of dress at home, is that the ongoing struggle for the right to the coastland lands in Bahia offers us a radical vision of what free Black life can look like. As Andrew W. Kahrl also writes, "Radical visions of a beloved community took place on the beach, while for others, the sands beneath their feet inspired dreams of abundance and personal liberation from the burden of race."[25] Like Kahrl's story of the African American history of beach land ownership, I hope this chapter has shed light on what he calls the "ties that bind race, power, and pleasure to the land."

NOTES

1   bell hooks, "Homeplace (A Site of Resistance)," in *Yearning: Race, Gender, and Cultural Politics* (Boston: South End Press, 1990).

2   Christen Smith, *Afro-Paradise: Blackness, Violence and Performance in Brazil* (Urbana: University of Illinois Press, 2016).

3   David Hellwig, *African-American Perspectives on Brazil Racial Paradise* (Philadelphia: Temple University Press, 1992).

4   Jeffrey Lesser, *Immigration, Ethnicity, and National Identity in Brazil, 1808 to the Present* (Cambridge: Cambridge University Press, 2013), 143.

5   Marcos Chor Maio, "Gilberto Freyre and the UNESCO Research Project on Race Relations in Brazil," in *Luso-Tropicalism and Its Discontents: The Making and Unmaking and Racial Exceptionalism*, ed. Warwick Anderson, Ricardo Roque, and Ricardo Ventura Santos (New York: Berghahn Books), 112–34.

6   See Cidinha da Silva, *#Parem de Nos Matar!* (São Paulo: Pólen, 2019); Marielle Franco, *UPP: A Redução da Favela a Três Letras: Uma análise da política de segurança pública do estado do Rio de Janeiro* (Rio de Janeiro: N-1 Edições, 2018); Lélia Gonzalez, *Primavera para as rosas negras: Lélia Gonzalez em primeira pessoa* (São Paulo: Editora Filhos da África, 2018); Sales Augusto dos Santos, *O Sistema de Cotas Para Negros da UnB: Um balanço da primeira geração* (Jundiaí: Paco Editorial, 2015).

7   Smith, *Afro-Paradise*, 5.

8   Kia Lilly Caldwell, *Health Equity in Brazil: Intersections of Gender, Race, and Policy* (New Brunswick, NJ: Rutgers University Press, 2017); Sonia E. Alvarez

and Kia Lilly Caldwell, "Promoting Feminist Americanidade: Bridging Black Feminist Cultures and Politics in the Americas," *Meridians* 14, no. 1 (2016): v–xi; Okezi Otovo, *Progressive Mothers, Better Babies: Race, Public Health, and the State in Brazil, 1850–1945* (Austin: University of Texas Press, 2016).

9 Jessica Lynn Graham, *Shifting the Meaning of Democracy: Race, Politics and Culture in the United States and Brazil* (Berkeley: University of California Press, 2019); Tianna Paschel, *Becoming Black Political Subjects: Movements and Ethno-Racial Rights in Colombia and Brazil* (Princeton, NJ: Princeton University Press, 2018); Watufani Poe, "Black Gay Worldmaking of the Global 1980s: Brazil and the United States," *Revista Brasileira de Estudos da Homocultura* 5, no. 18 (2023): 252–82.

10 Andrew W. Kahrl, *The Land Was Ours: How Black Beaches Became White Wealth in the Coastal South* (Chapel Hill: University of North Carolina Press, 2012), 6.

11 Kahrl, *The Land Was Ours*, 7.

12 Katia Costa Santos, "Black Girls in Ipanema" *NACLA Report on the Americas* 54, no. 3 (2022): 308.

13 Santos, 306.

14 Bill Donahue, "The Price of Gold: Inside the Troubling Legacy of Displacing Poor Communities for the Olympic Games—and One Village's Resistance in Brazil," *Washington Post*, July 6, 2020, www.washingtonpost.com.

15 Santos, "Black Girls in Ipanema," 303.

16 Smith, *Afro-Paradise*, 6.

17 Scott Ickes and Bernd Reiter, *The Making of Brazil's Black Mecca: Bahia Reconsidered* (East Lansing: Michigan State University Press, 2018).

18 Bryce Henson, *Emergent Quilombos: Black Life and Hip-Hop in Brazil* (Austin: University of Texas Press, 2024).

19 Keisha-Khan Y. Perry, "Housing Justice in the Americas: Struggle and Solidarity," *NACLA Report on the Americas: Fighting for Housing Justice in the Americas* 54, no. 3 (2022): 237.

20 Kevin Quashie, *Black Aliveness, or a Poetics of Being* (Durham, NC: Duke University Press, 2021); Santos, "Black Girls in Ipanema."

21 hooks, "Homeplace (A Site of Resistance)."

22 Isabel Wilkerson, *The Warmth of Other Suns: The Epic Story of America's Great Migration* (New York: Penguin Random House, 2010).

23 Courtney Morris, *To Defend This Sunrise: Black Women's Activism and the Authoritarian Turn in Nicaragua* (New Brunswick, NJ: Rutgers University Press, 2023).

24 Geri Augusto, "For Marielle: Mulhere(s) da Maré—Danger, Seeds and Tides," *Transition* 129 (2020): 250.

25 Kahrl, *The Land Was Ours*, 19.

PART IV

Developing the Beach

Tourism, Activism, and Regulation

Evening Shoreline (Setha Low, 2023)

13

Las Playas son del Pueblo

The Struggle for Puerto Rico's Coast

KATHERINE T. MCCAFFREY

Esta es mi playa
Este es mi sol
Esta es mi tierra
Esta soy yo
[This is my beach. This is my sun. This is my land. This is me.]
—"El Apagón-Aquí Vive Gente," Bad Bunny

María Rosado, born in 1932, grew up on the craggy north coast of Vieques Island, scampering along the rocks, collecting snails and crabs, swimming in the warm Atlantic waters with her siblings and other neighborhood children. She remembered the little beach, *La Lanchita*, a sandy strip between two cliffs, where neighborhood children would swim and men would go to sea in small wooden boats. The fish the men caught—snapper, dorado, grouper, and trunkfish— would be shared with family and neighbors. No one had refrigerators in those days. The fish would be consumed or dried, salted, smoked, and stored in barrels. It contributed to sustenance during the dead-time of the sugarcane season, when there was no work and wages to buy food.

Now *La Lanchita* is mostly gone, consumed by an expansive guest-house, *Casa La Lanchita*—a three-story terraced villa built into the rock over the little beach. With an infinity pool floating over the sea, a carport, and unobstructed views of the ocean, *Casa La Lanchita* is marketed by its owners as an upscale Caribbean retreat. *La Lanchita*, the beach of María's childhood, a place of sharing and drawing sustenance from the sea, has been swallowed by cement.

Since I first visited Vieques in 1991, I have watched a ragtag series of half-built, half-destroyed, sometimes-rented vacation homes mushroom along the north coast of the island. The houses are the conflicted legacy of a squatting movement that reclaimed living room from a military-controlled buffer zone. Between 1941 and 2003, the US Navy occupied the majority of Vieques Island, squeezing a civilian population of approximately ten thousand people into the center of the island. Wedged between an ammunition storage area and a live training range, residents lived in a state of limbo as economic stagnation and military antagonism threatened to displace them from their homes. In response to housing pressures and insecurity in the 1970s and 1980s, residents crossed into the outskirts of undeveloped land officially reserved for Navy maneuvers. They set up wooden stakes and rope to cordon off land in a series of organized "land rescues" and built *casitas* and eventually cement houses on land without title. Over time, this land opened by working-class residents seeking living room was penetrated by speculators seeking cheap beach front property. Today, the beach front is as an important site of struggle as working-class Vieques residents seek a foothold on an island rapidly gentrifying after a mass mobilization drove the Navy off the island in 2003.

Vieques's struggle against gentrification and displacement is shared by Puerto Rico as a whole after a turbulent decade defined by a debilitating debt crisis and a federally imposed austerity plan that decimated public services including health care, pensions, and education. A shrinking public sector and rising unemployment triggered a mass exodus of people. When two catastrophic hurricanes—Irma and Maria—hit back-to-back in 2017, flattening communities, crippling the electrical grid, and causing billions in damage, the population exodus accelerated. In the face of these relentless pressures, and ongoing government mismanagement and corruption, Puerto Rican citizens have demonstrated a remarkable ability to harness fury, organize, and challenge injustice, most dramatically in massive street protests in 2019 that forced the resignation of Governor Ricardo Rosselló.[1]

In 2022, Puerto Rican superstar rapper Bad Bunny released "El Apagón: Aquí Vive Gente" (The Power Outage: People Live Here), a fiercely political music video that extends a protest song into an eighteen-minute documentary denouncing the suffering and displacement of Puerto Ricans. The song ends with an affirmation of the centrality of the beach to Puerto Rican identity, the subject of this essay.

Puerto Rico's beaches are an ongoing focus of a multipronged fight for popular sovereignty and a sustainable future. On an island nation ringed by eight hundred miles of shore, community and environmental struggles over coastal resources in Puerto Rico are intensifying. The struggle over coastal resources, development, and access to the shore reveals the intersection of Puerto Rican history, culture, and political economy. First, I will briefly describe the historical understanding of the coast and coastal resources on the island—connected to a moral economy embedded in a modern capitalist plantation economy—which animates current clashes. A developmentalist approach post-WWII that envisioned the beach as a site of leisure, particularly for a foreign elite, stimulated the earliest protests against privatization of the beach in the 1960s. I will then highlight several ongoing battles over the beach that point to this arena as an ongoing site for competing views over Puerto Rico's sovereignty and future.

From Margin to Center

Until the mid-twentieth century, Puerto Rico's coastline was marginal to capital accumulation, and its land was devalued. The landscape of Old San Juan embodies these ideas. Just past El Morro, the sixteenth century Spanish fort protecting the valuable land, people, and resources of the urban center lies La Perla, Puerto Rico's oldest and most maligned shanty town. The Spanish sited a slaughterhouse here in the eighteenth century. Later, La Perla's land, outside the walls of the city on a peninsula's cliff, became home to the impoverished, including freed slaves who were not permitted to live in the city. In the 1960s, Oscar Lewis developed theories of a "culture of poverty" among its residents, who were trapped, he theorized, by generational poverty and developed attitudes and behaviors that reflected and reinforced their difficult life circumstances.[2] Next to La Perla, a series of densely packed white marble tombs and statuary define the Santa María Magdalena de Passis cemetery. Situating the cemetery on the waterfront on the outskirts of the city reflected the perception of the coast's marginal status in the nineteenth century, when concern for public health shaped efforts to move grave sites and disease away from population centers.[3]

The coast, however, and coastal resources have long played an important role in the sustenance, autonomy, and pleasure of working people, what theorists refer to as a "moral economy" that shapes people's ethical understandings of how economies should operate.[4] The commodification of the coast into "beach"—empty spaces of leisure—in the twentieth century erases diverse and productive coastal and ecological realities of the shore. Puerto Rico's coast encompasses shorelines that are both sandy and rocky, with bays and cays, coral reefs, shoals, lagoons, seagrass beds, mangrove forests, and coconut groves. Historically, these spaces supported different forms of flora and fauna: snails, crabs, diverse fish and shellfish, coconut, and wood. They also sustained diverse peoples. Juan Giusti Cordero's rich historical research in Piñones, Loiza, a sixteen-kilometer littoral zone east of San Juan, points to a millennium of occupation and coastal production: Tainos who cultivated manioc, yams, and maize in the rich soil of the barrier islands; Tainos and African maroons who fled enslavement in mines and haciendas to shelter in Piñones' dense mangrove forest; settlers, hunters, fugitives and outlaws who lived on fishing, contraband, and theft.[5] All of these varied ecosystems on the coast supported diverse people and their productive activities at different moments in time. In Sidney Mintz's classic life history *Worker in the Cane*, the protagonist, Don Taso Zayas Alvarado, vividly describes the shore as a place of abundance and joyful refuge from the brutal conditions of the sugar cane plantation.

> After we would work through the day, we used to go to the sea, to the part we call El Cayo [the Cay]. And we used to go with a person who could cast a throw-net. From here in the barrio we would carry the viandas [starchy root vegetables such as cassava, taro, and yam] They would get the firewood there; they would get rocks—those rocks from the cay there. They would make a fireplace, and on it they would put a five-gallon tin (*latón*), and they would peel the viandas, and they would put them on to boil, and we would go to cast the throw-net, and there we used to catch fish—lobster and all kinds of fish. You caught *jarea*, you caught *robalo*, you caught *jayao*, you caught *mero*—different kinds of fish. But always there were plenty of *jarea* above all. . . . And we went to work and prepared all those fish, and put them in a fish can to cook; and the viandas were already cooked, and we ate what we could . . . we ate fish until we were stuffed.[6]

More recently, Carlos G. García-Quijano and colleagues describe contemporary coastal resource foraging (CRF) in southeastern Puerto Rico—encompassing small-scale commercial fishing, recreational fishing for food, subsistence fishing, land crabbing, and oystering—as a distinct mode of production that contributes to overall personal satisfaction, work, and enjoyment of community life.[7] These findings dovetail with the research of David Griffith, Manuel Valdés Pizzini, and Jeffrey Johnson, which describes how Puerto Rican working-class men embrace fishing as a form of "therapy" that contrasts with wage labor, which is a source of injury and sickness.[8] García-Quijano and colleagues describe the emotional pull of CRF and how even individuals who have migrated to industrial wage labor strategize a return to their natal fishing communities.[9]

As common property, the Puerto Rican coast historically has provided access to natural resources that were not owned or managed by external agents. It offered space that existed outside of capitalist wage labor and exploitation and sustained community and patterns of reciprocity. When Don Taso described a *serenata* (an evening's enjoyment or meal) he emphasized that these gatherings were not primarily about physical sustenance, as cane workers would first eat dinner at home before going to the shore. Moreover, the *serenatas* were certainly not about earning money:

> We liked the night fishing so much—catching the fish—and we did it as a sport and for enjoyment at the same time. We never did it in order to sell the fish. When we held one of those *serenatas* it was to eat the fish there that way; if it was a big fish then people would bring back a couple of pounds of fish to their homes, but we never did it with the end in mind of selling the fish. In recent years people would go especially to catch lobster and such and sell them; but when it was for a *serenata* it was exclusively to eat, a matter of a fiesta and such.[10]

In my own ethnographic research in Vieques, I became interested in the strength of emotional connection to the coast and how, in the late 1970s, residents mobilized opposition to US Naval bombing exercises by rallying behind fishermen.[11] Fishermen framed collective grievances against expanded naval maneuvers and bombing exercises by focusing on reduced access to fishing waters during maneuvers and destruction of traps by ship traffic. Fishermen built popular support not by condemning US colonialism, or opposing military maneuvers or even bombing, but by demanding access to coastal resources.

The valorization of the beach as a place of sharing, gathering, and celebrating a communal identity continued in the more recent Vieques mobilization of 1999, when activists occupied the naval bombing zone on the eastern tip of the island, establishing a series of encampments on its white sand beaches. Fishermen ferried supporters out to the protest camps on a daily basis, their small skiffs laden with coolers of food and drink. There on the beach, at a makeshift kitchen under a tarp, women would fry fish and arepas for visitors, and ladle out plates of rice and beans.

The theme of access to natural resources continues to animate Vieques's struggle for rights to former base land now under control of the US Department of Fish and Wildlife. Residents bristle at fences, restrictions on access, and threats of fines for crabbing.[12] Their indignation reveals the power of a moral economy forged outside the capitalist wage economy and built on reciprocity and connection to the rich marine environment.

Commodification of the Shore

The conception of the Caribbean as an idyllic space of bright sun, white sand beaches, and turquoise water emerged in the mid-twentieth century with the rise of international leisure travel. The advent of affordable airline travel, extended global communications, and an integrated capitalist world economy reshaped ways of life.[13] The shoreline, which historically had represented a place of refuge and sustenance to marginalized peoples, was now a source of wealth creation. A Caribbean tourism emerged based on its natural resources—sun, sand, and sea—and was marketed as "paradise."[14] Today, the Caribbean is the most tourist dependent region in the world and several states, Antigua and Barbuda and Aruba, receive roughly 60 percent of their gross domestic product from tourism. Between 1995 and 2014, the volume of tourists in the region more than doubled from twelve million to twenty-six million.[15]

In comparison to more tourism dependent islands, Puerto Rico's tourism industry developed in the postwar period as one prong of a multifaceted development scheme that emphasized light manufacturing. "Operation Bootstrap," Puerto Rico's dependent economic development program, lured US corporations with a readymade industrial infrastructure and a ten-year exemption of corporate profits from both island and federal taxes.[16] Tourism, under theLuis Muñoz administration, worked to promote Puerto Rico as both a glamorous vacation destination and a modern state worthy of financial investment. Opening a tourist bureau in Rockefeller Center in the 1950s, the Muñoz administration targeted affluent East Coasters, conceiving of these tourists as potential future investors in the island.[17] The $7.2 million Caribe Hilton, an entirely state-financed hotel venture, crystallized Puerto Rico's commitment to showcasing the island as modern and dynamic, "a new Miami with a Spanish accent."[18]

By the 1960s, Puerto Rico's dependent economic model came under scrutiny of a nascent environmental movement, critical of infrastructure and industrial development that damaged coastal ecosystems, strip mined in the island's interior, and cordoned the beach from public access.[19] Carmen M. Concepción traces the earliest expressions of modern Puerto Rican environmental activism to the journalistic work of the esteemed novelist Enrique Laguerre, who criticized the Popular Democratic Party government's development strategy and its impact on the environment. Concepción notes that "Laguerre's strong opposition to the privatization of coastal areas for tourist development, which was one of his earliest clashes with the Puerto Rican government, led him to develop a campaign on the theme 'Beaches are public property,' from the late 1950s."[20] As tourism expanded in the 1960s, the encroachment of big hotels on beaches in the San Juan area, developing a zone for outsiders, triggered protest and resentment.

Contemporary Grassroots Struggles over the Coast

Here, I will highlight a series of coastal struggles and confrontations that have played out in Puerto Rico over the past twenty-five years. The movements have their roots in an older Puerto Rican environmental movement that historically has been militant and nationalistic. Compared to US mainstream environmentalism, the Puerto Rican environmental movement, like much of Latin American environmentalism, meshes ecological concerns to broader popular mobilizations for social justice and equity.[21] Concepción argues that from its earliest efforts to oppose mining, Puerto Rican environmental activism encompassed economic and political aspects, emphasized community organizing, and recognized health issues in both the workplace and community. In rallying around environmental concerns, she argues that the Left recognized "another space for struggle and for popular mobilization."[22]

With weak state enforcement of existing laws and troubling levels of corruption, grassroots mobilization remains a principal check on private appropriation of natural resources.[23] Movements draw strength from a national identity that defines beach access as part of the common property of the nation, an entitlement first defined by Spanish law in 1866 and enshrined in 1979 in Puerto Rico's Coastal Zone Management Program, which requires developers to retain public access to the shore. The movements draw energy, personnel, and tactical lessons from the battles against privatization that emerged in Puerto Rico in the 1990s, as the pro-statehood government of Pedro Rosselló privatized all government-run entities, including water, telephone, health care, and insurance. Struggles over the coast, like the successful Vieques mobilization that halted US Navy bombing exercises in 2003, incorporate diverse actors, civil disobedience, and direct action in coalitions that embrace grassroots struggles in places like Rincón, Isabela, Vega Baja, San Juan, and Vieques. While essentially a Puerto Rican based grassroots movement, mobilization for beach access integrates and draws on the expertise of international organizations such as the Sierra Club and Surfrider. Moreover, in recent years, especially since Hurricane Maria, struggles over the coast raise larger questions about popular sovereignty and survival as rising seas and warming oceans threaten the viability of coastal ecosystems.

Playas Pa'l Pueblo

As one of the most densely populated and heavily industrialized parts of metropolitan San Juan, Carolina saw its beach front systematically usurped in the 1950s and 1960s. The construction of an international airport, hotels and upscale housing in the postwar decades swallowed beaches that were once enjoyed by working-class residents. In the 1970s, episodes of civil disobedience under the banner of "Las Playas son del

Pueblo" (The beaches belong to the people) successfully challenged encroachment and privatization of the beaches in Carolina. Between 2005 and 2019, activists locked horns with Marriott International over the development of the Carolina coast, occupying the beach for fourteen years as part of a coalition movement opposed to privatization and environmental destruction.

In early 2005, the Courtyard Marriott secured approval from municipal and commonwealth agencies and politicians to build condos and a hotel on ecologically sensitive public land. Civil disobedience forced Marriott to back off its plans. Environmental activist Alberto de Jesus Mercado, better known as "Tito Kayak," and his direct action group, Amig@s del Mar, camped out at Carolina public beach to fight the proposed Marriott development. (Kayak is well-known for his highly dramatic, media-savvy protests. In November 2000, as part of protests against bombing in Vieques, Kayak unfurled a Puerto Rican flag from the crown of the Statue of Liberty and a banner declaring "Bieke [Vieques] o Muerte." This incident landed him in federal prison for a year.)

In Carolina, protesters dismantled iron rods and other building materials from the construction site. Kayak chained himself to a truck and blocked ongoing construction. Activists were arrested but continued to press for free access to the beach, establishing an encampment that they occupied for fourteen years, rotating personnel to maintain a constant presence.[24] Direct action and public attention spurred the Puerto Rican legislature to investigate how Marriott International had received permission to develop beachfront land. Protest also inspired residents to file a lawsuit against Marriott International, the Puerto Rico Environmental Quality Board, and other government agencies, which resulted in the annulment of Marriot's construction contract. In a victory for environmental groups, the decision put an end to Marriott's condo and hotel construction on public land.

Conflict between the municipality of Carolina and Marriott over rights to coastal land abutting the public beach area, continued, however, for over a decade until the parties finally reached an agreement in 2019. The municipal government paid US$2 million to recover four acres of beachfront land from Marriott, which it integrated into the existing public beach. Media coverage indicated that demonstrators and environmental movement leaders supported the resolution which would end a protracted battle and free the beach for public use.[25] As part of the resolution, the protest encampment was dismantled. In reality, the "ecological reserve" that environmental activists cultivated over fourteen years, with carefully selected native flora, was razed by the municipality to recreate a 1950s era empty, white sand, coconut palm-lined beach. "They are dismantling on the littoral coast, the ecological forest of the northeast," lamented one activist as chainsaws rumbled in the background. "It is a little lung of our littoral coast."[26] Such comments reflected a subtle shift in the understanding of Puerto Rico's coast as an ecologically fragile habitat and essential defense against ferocious storms and rising seas.

Isleta de San Juan

Over the past twenty years, two major battles over coastal access have unfolded on the narrow neck of a peninsula, Isleta de San Juan, that leads to historic Old San Juan. While the first struggle—centered on opposition to the Paseo Caribe complex—failed to halt the construction of a mega luxury residential and commercial complex, it did launch a wave of mobilizations to claim coastal access and limit development in the early aughts. The second ongoing struggle in the shadows of Paseo Caribe, Escambrón Unido, challenges the usurpation of a public beach and park as part of a redevelopment of the historic Normandie Hotel.

The Sit-In at Paseo Caribe

Paseo Caribe was a US$300 million development site on the peninsula of Condado Bay in San Juan. At the very tip of the Condado Peninsula sits a small seventeenth century fort, San Gerónimo, built by the Spanish to protect the entrance to the bay. Paseo Caribe would block access to the fort and the peninsular coast. In July 2007, five activists from the organization Amig@s del Mar, led by Tito Kayak, climbed up and occupied four giant cranes at the construction site. They were protesting the government, developer, and politicians for privatizing public lands and denying the people access to the fort and the beach. Kayak remained perched on the crane for one week. By the time police moved in to arrest him, thousands of supporters had gathered, eating, playing music, and blocking traffic. The confrontation between Kayak and the police is best described by Tito's own Wikipedia page: "In a daring escape, he rappelled down from the crane and unto [*sic*] a red kayak in the water below while police officers were kept at bay by his supporters. Tito then rowed himself under a bridge whose clearance was too low for the police powerboats and switched out of the kayak, so when the kayak was apprehended he was no longer on board. Instead, he swam across to the other shore. When he was spotted by the police helicopters, supporters jumped into the water confusing the police further and finally guaranteeing Tito's getaway."[27] Photos of Kayak enjoying a sandwich and a glass of red wine surfaced later that evening on the internet.

The Amig@s del Mar action launched a campaign that continued for the rest of the year, in which dozens of activists, actors, and artists staged sit-down protests at the entrance of the Paseo Caribe development site, demanding the halt of construction and the demolition of the luxury complex.

The developer, San Gerónimo Caribe Project, ultimately prevailed in the construction battle. Although the Puerto Rican government, taken off guard by the scale of public opposition, temporarily halted the project, a subsequent Supreme Court ruling reversed the decision, allowing the construction to resume. Alluding to the battle to build the hotel, the lead architectural firm claimed that the "long wait was worth it": "Paseo Caribe is a testament to overcoming cultural, political and technical challenges to create something great. This project, in fact, took over fifteen years to develop, en route to completing a memorable plan on this high profile site."[28] During that time period, the original developer withdrew from the project and prodevelopment forces attempted to criminalize activism at construction sites. An amendment to the Puerto Rican penal code, popularly known as "Ley Tito Kayak," made it a felony to enter a construction site for the purpose of obstructing work. In a victory for Tito Kayak, the law was later declared unconstitutional.[29]

Escambrón Unido

An ongoing battle over a waterfront public beach and park next to the site of the vacant landmark Normandie Hotel captures some of the current dynamics of struggle, which are increasingly animated by unbridled coastal development. In 2023, on the Isleta de San Juan alone, developers proposed an over US$1 billion investment in luxury housing and tourist development, pending financing and permits.[30] In a context where real estate rules, the threat of climate change and concern for democratic norms appear to be an afterthought.

Situated in the Escambrón area at the entrance to the Isleta de San Juan, the elegant art deco Normandie Hotel was a sensation when it debuted in the fall of 1942. Modeled after the French transatlantic passenger ship the SS *Normandie*, and one of the first luxury hotels on the island, its nautical silhouette evoked "movement, modernism and technology."[31] The Normandie was landmarked in 1980, and its devotees have described it as a "Boricua Taj Mahal" and an important part of national patrimony.[32] Its location, however, has always posed challenges to its viability; it was expropriated by the military during WWII and heavily damaged by Hurricane Georges in 1998, and, most concerning from a developer's viewpoint, it lacks in beach access and parking.

The hotel was shuttered in 2009, and Puerto Rico representative Angel Matos García introduced a measure to expropriate and demolish the Normandie in March 2021, which sparked push back from the Instituto de Cultura Puertorriqueña and Historic Commission.[33] Several engineers recommended destruction, but a New York based real estate firm, the Ishay Group, acquired the hotel for US$8.6 million in January 2022 under Act 60, a tax incentive that exempts businesses and investors who establish themselves in Puerto Rico. In addition, Ishay was promised federal benefits for "opportunity zones" and access to federal reconstruction funds. Ishay had plans to invest US$100 million in the Normandie's renovation.[34]

Yet while the abandoned hotel languished for over a decade, the adjacent oceanfront municipal land was "in use from dawn to dusk," explained community leader Gradiza Fernández. "From 6 AM there are people training, taking yoga classes, learning to swim, running on the track."[35] Indeed, for nearly one hundred years, land next to the Normandie, containing the Third Millennium Park and the Sixto Escobar Stadium, was zoned as a public park and beach and used for recreational activities and sports such as baseball, soccer, and surfing. And before it was a park, it was a swamp.[36] A 2021 hazard mitigation plan prepared by the Municipality of San Juan with FEMA funds identified the Escambrón sector as a high-risk area, vulnerable to flooding and storm surges.[37] With the sea level projected to rise six feet by 2050, a high water table will likely threaten the viability of this land.

As part of the Normandie development package, and without any hearings, public input, or environmental impact statement, the San Juan City Council sweetened the deal, leasing Ishay Group a section of beachfront public park to construct an underground parking lot with five hundred parking spaces and an event pavilion.[38] Residents were incensed by a lack of transparency in all levels of communication about the process. Fernández explained, "We saw very strange things—people digging holes in the ground." Residents learned in the press that the developer planned to build a parking lot on Sixto Escobar park and rallied to defend public land. "When we heard they were planning a parking garage, we organized all the people who love this area to defend what little green space we have left, a free space for exercise."[39] For over a year, protesters camped out at the entrance of Escambrón, held daily vigils, and marched down the streets of Old San Juan with hundreds of supporters, beating pots and pans.[40] They organized swim-ins to claim public access to the coastal waters and maintained an active social media presence. They spoke out against development that sacrificed the public interest to private wealth creation. Pedro Cardona Roig, a respected opponent and outspoken architect, urbanist, and planner who maintains a high-profile social media presence, summarized the stakes: "You cannot convert a public park into a parking lot without a process of public notice and hearings, without sharing a plan, without community participation. The Mayor is turning his back on San Juan residents and all citizens. This park belongs to all Puerto Ricans, more than to the municipality of San Juan or San Juan residents. The Mayor is the custodian, but he does not own this property. It belongs to the people of Puerto Rico. Our demand is for transparency, compliance with the law, that this be an open process."[41] The struggle over Escambrón Beach thus reveals the way conflict over public access to the beach and the protection of the coast from unbridled development is fundamentally about democracy and national sovereignty. In the aftermath of Hurricane Maria, as the Puerto Rican population has fled increasingly desperate conditions created by austerity and magnified by environmental destruction, the economic restructuring plan imposed upon the island has stifled growth and failed to deliver economic stability.[42] US-based firms and vulture capitalists have descended upon the island.[43] Rather than stimulating growth, the profits from capital investments flow away from Puerto Rico; for example, in the aftermath of Hurricanes Irene and Maria, rather than bolstering a devasted economy, the overwhelming majority of billions of federal reconstruction funds went to US-based contractors.[44] The Escambrón development follows this pattern, delivering public land on a silver platter to a US-based developer through tax incentive program only available to external developers. As activists reject backroom deals, they work to publicize government documents and demand transparency.[45]

Yo Soy Carey

In July 2021 an endangered hawksbill sea turtle (*carey* in Spanish) entered a construction site on Los Almendros Beach in Rincón where a condo association was rebuilding a beachfront pool that had been destroyed by Hurricane Maria. After laying over a hundred eggs in the sand, the turtle became trapped behind construction barriers and fences and was unable to return to the sea until construction workers intervened. The plight of the turtle inspired a community-based protest movement demanding a halt to construction that impeded public access to the shore and destroyed the shrinking habitat of endangered turtles.

The trapped turtle revealed the urgency of defining the parameters of Puerto Rico's coast at a moment when rising seas are rapidly eroding the existing coastal zone. According to the United States Geological Survey, in the aftermath of Hurricanes Irma and Maria, Puerto Rico suffered extensive damage to beaches, dunes, and coral reefs.[46] Such erosion fueled the conflict with the condo association which initially received permits to rebuild its pool on a shrinking beach.

The plight of the endangered turtle resonated with besieged residents who have suffered years of relentless austerity measures while government tax incentive policies have courted millionaires to come invest in the island and presumably solve Puerto Rico's financial woes. Yet tax laws attracting bitcoin millionaires have caused exorbitant hikes in rent island-wide, threatening modest family homeownership. One vocal critic, Alexandra-Marie Figueroa of Taller Salud, a Puerto Rican feminist group, argued that these tax laws were "crafted behind closed doors to make Puerto Rico a paradise for those that barge in and can afford it, but a life-sentence for those of us who try to hold on to what we have left of our country."[47] In this context, a turtle's struggle to maintain its existence provoked a powerful community response.

Defenders of the turtle, and, more broadly, access to the beach, demanded a halt to the construction. They were supported initially by the secretary of the Department of Natural Resources, who was subsequently pressured to reverse his decision. Protesters established a lively encampment at the construction site, declaring "Yo soy Carey" (I am the hawksbill turtle). They were met with tactical squads of police in riot gear; in one instance two hundred police in helmets wielding batons set upon ten sleeping young women before daybreak.[48] The disproportionate police response inspired broad community support and funds for the encampment. Thousands of people participated in demonstrations, contributed food and water, and purchased merchandise online to support the protest.

The court eventually ruled that the pool construction was illegal and ordered construction to halt and the condo association to dismantle the existing construction. Months went by and the condo association did not comply and the court did not enforce its order. Protesters took action. In March 2022, hundreds of demonstrators from all over Puerto Rico arrived in Rincón with sledgehammers in a pop-up, grassroots demolition effort.

Conclusion: Climate and the Future of the Coast

According to the National Oceanic and Atmospheric Administration, out of a total population of 3.6 million people, two-thirds of Puerto Ricans (2.4 million) live on coastal portions of the island.[49] The Climate Change Council of Puerto Rico notes that "all of Puerto Rico's major infrastructure—airports, power plants, thousands of miles of electrical lines, communications, aqueducts, sewers and ports facilities are located within 1 km of the shoreline."[50] And while scientists recognize that a high concentration of population and critical infrastructure in low-lying coastal areas magnifies the effects of coastal flooding and beach erosion and increases vulnerability to sea-level rise and storm surge, this environmental reality clashes with a development ideology and political economy that incentivizes coastal development.[51] While the consequences of coastal erosion are highly visible and concerning to residents,[52] and a panel of scientific advisors urged the Puerto Rican government to adopt a moratorium on coastal construction, the state has been slow to respond. In 2021, the first year of the Pedro Pierluisi administration, the government approved nearly 30 percent more coastal construction permits than in the previous year.[53] In April 2023, the Puerto Rican government declared a state of emergency to fight worsening coastal erosion. However, it has yet to declare a moratorium on coastal development.

Environmental struggles in Puerto Rico, as in the rest of Latin America, are inseparable from broader issues of sovereignty and self-determination. Puerto Rico's core conflict stems from its status as a US colony. While Puerto Rican party politics are paralyzed over status issues, Puerto Rican community, labor, environmental, and antimilitary movements—often overlapping and mutually reinforcing—become the avenue through which Puerto Ricans fundamentally engage the issue of sovereignty. The struggle over the coast is a vibrant expression of this fight. But all these popular movements, to quote the University of Puerto Rico sociologist Rafael Bernabe, share one characteristic: "They all embody the struggle of people to more directly control their lives, from their work conditions, the quality of the environment, the way their communities are policed, or how the state budget is distributed. They are all, in that sense, struggles for self-determination."[54]

NOTES

1 For English language news coverage of the epic protests see, for example, "Street Protests Shook the Halls of Power in Puerto Rico," *New York Times*, July 27, 2019, 12, www.nytimes.com; Lucinda Elliot, "Protests Rock Puerto Rico after Leaders Insult Hurricane Victims," *The Times*, July 24, 2019, 30, www.thetimes.co.uk. An anthem for the mobilization, "Afilando Los Cuchillos" ("Sharpening the Knives") by musicians Bad Bunny, Residente, and iLe captured the fury of the moment. See "Residente, iLe & Bad Bunny - Afilando los Cuchillos," YouTube Video, 00:05:19, July 17, 2019, https://www.youtube.com/watch?v=RSh7HIH2pvg. For analysis of the larger context of protest, see Marisol LeBrón, "The Protests in Puerto Rico are About Life and Death," *NACLA Report on the Americas*, July 18, 2019, https://nacla.org.

2 Oscar Lewis, *La Vida: A Puerto Rican Family in the Culture of Poverty—San Juan and New York* (New York: Random House, 1966).

3 Juan Carlo García-Cacho, "El Cementerio Santa María Magdalena de Pazzis, 1814–1899: Un Acercamiento de la Historia Cultural" (PhD diss., Universidad de Puerto Rico, 2012).

4   Marc Edelman, "Bringing the Moral Economy Back in to the Study of 21st-century Transnational Peasant Movements," *American Anthropologist* 107, no. 3 (2005): 331–45; Cynthia Grace-McCaskey, David Griffith, H. Lloréns, Carlos García-Quijano, and Miguel Del Pozo, "Negotiating Political and Moral Econo-mies in the US Caribbean After Hurricanes Irma and María," *Caribbean Studies* 49, no. 1 (2021): 3–27.

5   Juan Giusti-Cordero, "Labour, Ecology and History in a Puerto Rican Plantation Region: 'Classic' Rural Proletarians Revisited," *International Review of Social His-tory* 41 (December 1996): 53–82.

6   Sidney W. Mintz, *Worker in the Cane: A Puerto Rican Life History* (New York: W. W. Norton, 1974): 61.

7   Carlos G. García-Quijano, John J. Poggie, Ana Pitchon, and Miguel H. Del Pozo, "Coastal Resource Foraging, Life Satisfaction, and Well-Being in Southern Puerto Rico," *Journal of Anthropological Research* 71, no. 2 (2015): 145–67.

8   David Griffith, Manuel Valdés Pizzini, and Jeffrey C. Johnson, "Injury and Ther-apy: Proletarianization in Puerto Rico's Fisheries," *American Ethnologist* 19, no. 1 (1992): 53–74, https://doi.org/10.1525/ae.1992.19.1.02a00040; David Griffith and Manuel Valdés Pizzini, *Fishers at Work, Workers at Sea: A Puerto Rican Journey Through Labor and Refuge* (Philadelphia, PA: Temple University Press, 2011).

9   García-Quijano et al., "Coastal Resource Foraging."

10 Mintz, *Worker in the Cane.*

11 Katherine T. McCaffrey, *Military Power and Popular Protest: The U.S. Navy in Vieques, Puerto Rico* (New Brunswick, NJ: Rutgers University Press, 2002).

12 Katherine T. McCaffrey, "Environmental Remediation and Its Discontents: The Contested Cleanup of Vieques, Puerto Rico," *Journal of Political Ecology* 25, no. 1 (2018): 80–103. https://doi.org/10.2458/v25i1.22631.

13 Dennis Merrill, "Negotiating Cold War Paradise: U.S. Tourism, Economic Plan-ning, and Cultural Modernity in Twentieth-Century Puerto Rico," *Diplomatic History* 25, no. 2 (2001): 179–214.

14 Marian A. L. Miller, "Paradise Sold, Paradise Lost: Jamaica's Environment and Culture in the Tourism Marketplace," in *Beyond Sun and Sand*, ed. Sherrie L. Baver and Barbara Deutsch Lynch (New Brunswick, NJ: Rutgers University Press, 2006).

15 Sebastian Acevedo Mejia, Trevor Serge Coleridge Alleyne, and Rafael Romeu, "Revisiting the Potential Impact to the Rest of the Caribbean from Opening US-Cuba Tourism," *IMF Working Papers* 2017, no. 100 (2017): A001. https://doi.org/10.5089/9781475595727.001.A001.

16 Sherrie L. Baver, *The Political Economy of Colonialism: The State and Industri-alization in Puerto Rico* (Westport, CT: Praeger, 1993); James L. Dietz, *Economic History of Puerto Rico: Institutional Change and Capitalist Development* (Princ-eton, NJ: Princeton University Press, 2018).

17 Merrill, "Negotiating Cold War Paradise."

18 Merrill.

19   Manuel Valdés Pizzini, "Historical Contentions and Future Trends in the Coastal Zones: The Environmental Movement in Puerto Rico," in *Beyond Sun and Sand*, ed. Sherrie L. Baver and Barbara Deutsch Lynch (New Brunswick, NJ: Rutgers University Press, 2006), 44–64.

20   Carmen M. Concepción, "The Origins of Modern Environmental Activism in Puerto Rico in the 1960s," *International Journal of Urban & Regional Research* 19, no. 1 (1995): 112.

21   David V. Carruthers, *Environmental Justice in Latin America: Problems, Promise, and Practice* (Cambridge, MA: MIT Press, 2008).

22   Concepción, "Origins of Modern Environmental Activism in Puerto Rico."

23   With considerable attention to the problem of corruption in Puerto Rico, I am also mindful to Villanueva's insightful work on the "contradictory 'political work' of corruption discourse, both in extending colonial work and in resisting it." Joaquín Villanueva, "Corruption Narratives and Colonial Technologies in Puerto Rico," *NACLA Report on the Americas*, 52, no. 2 (2019): 188–93. For the problem of corruption, see, for example Ricardo R. Fuentes-Ramírez and Julio Quintana Díaz, *El Costo Económico de la pobre gobernanza y la corrupción en Puerto Rico* (Ponce, Puerto Rico: Ponce, Puerto Rico, 2023); Amelia Cheatham and Diana Roy, "Puerto Rico: A US Territory in Crisis," *Council on Foreign Relations*, last updated September 29, 2022, www.cfr.org.

24   The following YouTube clips record activists discussing the movement, their participation, and finally, the dismantling of the encampment: Carlos Pérez, "Playas pal' Pueblo El Pueblo habla," YouTube video, 00:00:59, December 9, 2015, https://www.youtube.com/watch?v=DVAotS_5Ewo;Perez; Carlos Pérez, "Playas pal' Pueblo El Pueblo Habla 2," YouTube video, 00:00:23 December 9, 2015, https://www.youtube.com/watch?v=OkWwWzzYJLs; Defend Puerto Rico, "Playas pal' Pueblo | #DefendPR," YouTube video, 00:04:54, August 5, 2018, https://www.youtube.com/watch?v=4XUnPs5hXZg; José López, "Las playas son o no pal pueblo" YouTube video, 00:25:49 June 6, 2019, https://www.youtube.com/watch?v=oZgnI8AyR1g. Melissa Rosario describes the stalled quality of this protest movement that existed for years with an unclear path to victory. Melissa Rosario, "Inhabitating the Aporias of Empire: Protest Politics in Contemporary Puerto Rico," in *Ethnographies of U.S. Empire*, ed. Carole McGranahan and John F. Collins (Durham, NC: Duke University Press, 2018), 112–128.

25   "Carolina Municipal Government Retakes 4+ Acres of Beachfront after 10 Year Fight," *News is my Business*, April 23, 2019, https://newsismybusiness.com.

26   Jose Lopez, "Las playas son o no pal pueblo," YouTube video, 00:25:49, June 6, 2019, https://www.youtube.com/watch?v=oZgnI8AyR1g.

27   "Tito Kayak," *Wikipedia*, accessed August 28, 2023, https://en.wikipedia.org.

28   RSP Architects, *Paseo Caribe: Breathtaking Views Set a Stage for a Vibrant Experience*, accessed August 28, 2023, https://rsparch.com/project/paseo-caribe/.

29   "Tito Kayak," *Wikipedia*.

30 José Orlando Delgado Rivera , "Proyectos turísticos en la isleta de San Juan suman $1,000 millones en inversión," *El Nuevo Día*, February 18, 2023, www.elnuevodia.com.

31 Johnny Torres Rivera, "Hotel Normandie," Puerto Rico Historic Buildings Drawings Society, June 21, 2014, www.prhbds.org.

32 Susanne Ramírez de Arellano, "Puerto Rico's Normandie Hotel a Reflection of Colonialism," Latino Rebels, February 9, 2022, www.latinorebels.com.

33 Giovanna Garofalo, "ICP opposes demolition of the former Normandie Hotel," *News Journal*, March 18, 2021 www.theweeklyjournal.com.

34 Mivette Vega, "NY Entrepreneur Has Big Plans for Puerto Rico's Normandie Hotel," *Floricua*, February 16, 2023, https://theamericanonews.com.

35 Bonita Radio, "QPEN En defensa del Escambrón," YouTube video, 00:59:20, November 2, 2022, https://www.youtube.com/watch?v=pk34SebDhGA.

36 Aníbal Sepúlveda, "El parking en la ciénaga del Escambrón: ¡Protejamos la Capital de la codicia pantanosa!" *El Nuevo Día*, September 1, 2022, www.elnuevodia.com.

37 Junta de Planificacíon, *2021 Municipio de San Juan: Plan de mitigación contra peligros naturales* (San Juan, Puerto Rico: Junta de Planificacíon and Municipio Autónomo de San Juan, 2021).

38 Sandra Rodríguez Cotto, "Municipio aprueba demoler edificio del Sixto Escobar para crear un 'parking' para el Normandie," *Ey Boricua*, August 25, 2022, https://eyboricua.com.

39 Bonita Radio, "QPEN En defensa del Escambrón."

40 El País, "La lucha de Puerto Rico por arrebatarle las playas a la privatización," YouTube video, 00:02:03, Novemver 28, 2022, https://www.youtube.com/watch?v=nA6-irD1248.

41 Bonita Radio, "QPEN En defensa del Escambrón."

42 Deepak Lamba-Nieves, Sergio M. Marxuach, Rosanna Torres, *PROMESA: A Failed Colonial Experiment?* (San Juan, Puerto Rico: Center for a New Economy, 2021).

43 Yarimar Bonilla, "The Coloniality of Disaster: Race, Empire, and the Temporal Logics of Emergency in Puerto Rico, USA," Political Geography 78 (April 2020): 102181; Yarimar Bonilla, Marisol LeBrón, Sarah Molinari, Isabel Guzzardo Tamargo, and Kimberly Roa, "The Puerto Rico Syllabus," https://puertoricosyllabus.com; Yarimar Bonilla and Marisol LeBrón, *Aftershocks of Disaster: Puerto Rico Before and After the Storm* (Chicago: Haymarket Books, 2019); "PROMESA Has Failed: How a Colonial Board is Enriching Wall Street and Hurting Puerto Rico," Center for Popular Democracy, September 14, 2021, www.populardemocracy.org.

44 Sergio M. Marxuach, "Puerto Rico Recovery Task Force," Centro para uno Nueva Economía, September 28, 2022, https://grupocne.org.

45 Escuchar LaVerdad, "PUERTO RICO: Sigue la lucha comunitaria para proteger el Escambrón," YouTube video, 00:32:30, February 20, 2023, https://www.youtube.com/watch?v=Zm3CHUqBjp8.

46 Supplemental Appropriations for Disaster Recovery Activities, "Assessment of Coastal Hazards for Puerto Rico," USGS, accessed August 30, 2023, https://www. usgs.gov.

47 M. Espada, "Influencers, Developers, Crypto Currency Tycoons: How Puerto Ricans Are Fighting Back Against the Outsiders Using the Island as a Tax Haven" *Time*, last updated April 19, 2021, https://time.com/.

48 Carlos Tolentino Rosario, "Movilizan a cerca de 200 policías a la playa Los Almendros en Rincón," *El Nuevo Día*, August 3, 2021, www.elnuevodia.com.

49 NOAA, "Puerto Rico: Costal Management," accessed September 1, 2024, https:// coast.noaa.gov/states/puerto-rico.html.

50 Puerto Rico's State of the Climate 2014–2021. P.R. Climate Change Council. Department: P. R. C. Z. M. Program 2022 San Juan, Puerto Rico: Department of Natural and Environmental Resources, NOAA Office of Ocean and Coastal Resource Management, https://www.pr-ccc.org/wp-content/uploads/2019/04/ PR_StateOfTheClimate_2014-2021_PRCCC-09-2022.pdf.

51 D. R. Reidmiller, C. W. Avery, D. R. Easterling, K. E. Kunkel, K. L. M. Lewis, T. K. Maycock, and B. C. Stewart, eds., *Impacts, Risks, and Adaptation in the United States: Fourth National Climate Assessment, Volume II* (Washington, DC: US Global Change Research Program, 2018).

52 Bianca Graulau, "Sea Level Rise is Swallowing Puerto Rico's Beaches," YouTube video, 00:05:10, November 7, 2019, https://www.youtube.com/ watch?v=RdixQEweDco; Maritza Barreto-Orta, "El Estado de las playas de Puerto Rico Post María," December 7, 2022, https://storymaps.arcgis.com.

53 V. Rodríguez Velázquez, "Construction Permit Approvals for Coastal Projects Fast-Tracked During Pierluisi's First Year," January 28, 2022, www.latinorebels. com.

54 R. Bernabe, "Puerto Rico's New Era: A Crisis in Crisis Management," NACLA Report on the Americas, November 26, 2007, https://nacla.org/.

14

Dublin's Riviera

Dollymount Strand and Bull Island

PAUL ROUSE

James Joyce got to the heart of Dollymount Strand's contrasting aspects when he wrote in *Ulysses* of its "grainy sands" and its "sewage breath."[1] The strand runs for five kilometers, the full length of Bull Island, which averages some seven hundred meters in width and sits in Dublin Bay. The island is connected to the mainland at the point nearest the city by a short wooden bridge and at its midway point by a causeway. Dollymount's proximity to the city center has marked it as a vital place of leisure and recreation for the people of Dublin.

The popularization of Dollymount as a seaside destination for day-trippers from the middle of the nineteenth century was a process that happened in tandem with, and was influenced by, the making of the modern sporting world, with its networks of clubs, governing bodies, and dedicated sporting venues. From the construction of two private golf courses in its dunes to the provision of changing shelters for swimmers, Dollymount carries the physical traces of the evolution of sporting activities. The beach and its sea has also offered a venue for all manner of play, including children and adults creating their own imitative versions of sporting practices.

During the twentieth century, increased awareness of Bull Island's unique habitat and a changing approach to environmental matters added a further dimension. The dynamic ecological system of the island, with its salt and freshwater marshes, its dune system, and its sandy beach, demanded the management of practices by users of the island that have presented a significant challenge. Its remarkable range of birds—and its consequent attraction for bird watchers—has seen it acquire a pioneering status as Ireland's first bird sanctuary. This, too, has been reflected in public policy and in the manner in which sport and popular recreation are now considered in a new context of environmental sustainability.

This chapter will examine the tension inherent in a small area that seeks to be at once a place of popular recreation, a venue for modern sports, and a United Nations Educational, Scientific, and Cultural Organization (UNESCO) biosphere reserve, with important species of flora and fauna. It will explore how the use of Bull Island has evolved in the course of its history and assess how wider changes in social and cultural practices impacted the beach and its hinterland. Finally, the chapter will look at what ownership of the island and the usage of its land reveal about the nature of wealth and power in Dublin.

The Making of an Island and Its Beach

Bull Island was the byproduct of plans (drawn up by the maritime cartographer and later notorious ship commander, Captain William Bligh) to deepen the waters of Dublin Bay and deal with the sandbanks that were so hazardous to the expanding trade of Dublin Port in the late eighteenth century. Two walls were built, a South Bull Wall (1795) and a North Bull Wall (1824), and in the wake of their construction, Bull Island formed where sandbanks had previously risen, outside the North Bull Wall. The new island grew out into Dublin Bay, decade after decade, its shorelines expanding with the deposition of sand and other material transported on the tides and winds.

Almost immediately after its construction, the North Bull Wall and Bull Island became a popular place to swim. The notoriety gained through drowning tragedies—and there were many—did not appear to diminish the crowds. Indeed, those crowds grew with the extension of the tram system to the wooden bridge connected to the island in the 1880s and later with the provision of buses. In the popular culture of inner-city Dublin, Dollymount Strand was the seaside, commonly known as "Dollier." As the geography of the city expanded and the coastal land parallel to Bull Island was swallowed by prosperous suburbs, popular use of the strand developed still further: Dollymount Strand was thronged with people of every class, drawn to the city island and its beach.

The water and the sands were used by people seeking a day out, healthy exercise, and a sense of wilderness. Some came to bathe in the waters or to walk through the dunes or along the sand. Others were beachgoers who sought to spend time in relaxation; still more were bird watchers or nature enthusiasts who wished to spend time on a beach that was essentially undeveloped. As it grew in popularity in the nineteenth century, the island itself and its beach did not develop the rows of hotels, ice cream shops, and seaside amusements commonplace in the Victorian world; rather, those enterprises developed in a limited way along the seafront facing the island. And with the exception of the Great War, when troops were based there, Bull Island has never had many people living on it; the population reached its peak at forty-six in 1936, but in 2022, numbered fewer than twenty.

Sport on Bull Island

It also became a place for sport. The making of Bull Island coincided with the transformation of play from traditional to modern structures of sport in the middle decades of the nineteenth century. Horseracing was then the most advanced sport in Ireland; it had a calendar of events, professionals who made their living from its increasing commercialization, and a formal structure of organization. By 1850, horserace meetings on Dollymount Strand were a regular occurrence, with significant prizes on offer; for the summer races of that year, a huge crowd assembled and numerous tents were erected along the beach to accommodate their refreshment and gambling needs.[2]

Bull Island was also the venue for a landmark moment in the development of international sport in Ireland. In 1874, a team chosen by the Irish Rifle Association (formed in 1867) represented Ireland against America in New York. The following year, a return match was organized for Bull Island, where the Dublin Shooting Club had a range from at least the 1860s. The match was set for June 29, 1875, and what ensued was the first sporting event in Ireland that was an international spectacle. In the weeks before the contest, the newspapers wrote in breathless detail of the voyage of the Americans across the Atlantic, of their kissing of the Blarney Stone on arrival in Ireland, their free travel around Ireland and free access to all amenities, the crowds that cheered them as they passed in open-top procession through Dublin's streets, and the many receptions held in their honor.[3]

So great was the interest that the area at Dollymount Strand dunes was unable to safely hold the crowds. The scale of the crowd crossing the narrow wooden bridge between the island and the mainland was extremely dangerous: "The crush was positively terrific. Little children ran serious risk of being stifled and many ladies had to be extricated from the press by their male friends to save them from fainting."[4] Further, lines of spectators stood "so close to the line of aim" that the effect on the shooters was "trying" according to one of the Irish organizers.[5] The Americans won the contest, but what was most striking was the extent to which a new phenomenon was manifesting itself; a writer in the *Irish Times* commented in great wonder that there was no denying the extraordinary hold that the contest had taken on the popular imagination. On top of that, the involvement of rival rifle manufacturers with the opposing teams offered an early insight to the manner in which celebrity, commercialism, and competition would blend together in sport.[6]

There are two further points to note here. First, this was sport for a certain class. It is true that among the crowd were members of the expanding Irish middle classes and working-class Dubliners. But this was sport, in participation terms, as the preserve of the wealthy. Second, that a beach hosted such a major sporting event emphasized the extent to which modern sport was in its formative phase, without extensive dedicated venues capable of holding large crowds. This, too, would change in the decades that followed as the desire to charge entrance fees demanded purpose-built or, at the very least, enclosable facilities to which it was easy to deny unrestricted access.

Most of the organized sport that took place from the late nineteenth century on Dollymount Strand did not draw large crowds, however. Instead, and increasingly, it played host to sporting events that were part of the great phenomenal explosion of grassroots sporting activity that continued from the late nineteenth century. The waters off Dollymount Strand were used time and again by the sailing clubs of the north coast of Dublin.[7] There were also open-sea swimming competitions, including those run by the North Dublin Winter Swimming Club. The sport of sea angling was practiced at the end of the island that was furthest from the city center, where bass anglers (albeit counseled patience) used fresh ragworm and lugworm to catch sea bass.[8] Dollymount also hosted low-level soccer matches, hurling teams, and athletics competitions, to name just three. In 1907, for example, the Irish Vegetarian Society held an athletics meeting on the beach.[9] In this usage, sport both formal and informal accommodated itself on public space. The beach and the island were a place of popular recreation of which sport was a part but did not colonize land except while it was being used. That is to say, it was sport that was portable and conducted in a shared space.

Golf and the Privatization of Public Space

A new sporting world was in the making, however, which drove the creation of dedicated sporting facilities. The establishment of clubs and governing bodies, their scale and diversity, their burgeoning civic importance, and their centrality to social and cultural life had an increasingly significant impact on first urban and then rural living. Part of this impact was the transformation of the physical landscape through the construction of new sporting grounds. No piece of land was beyond consideration as sports clubs sought their own dedicated playing space, a focal point for whichever sport they were seeking to promote, and, ultimately, a location that would promise the sort of permanence that would facilitate prosperity.

This had long-term implications for Bull Island and Dollymount Strand. The establishment of golf courses along the coast of Scotland from the mid-eighteenth century had proven a catalyst for the establishment of that sport. When it spread into Ireland in the 1880s, enthusiasts for golf sought to establish a club in Dublin city. The Dublin Golf Club was founded in 1885 by John Lumsden, a bank manager, who laid down a golf course at Phoenix Park in central Dublin. The club, whose membership comprised prosperous businessmen, landowners, British army officers, and Trinity College Dublin students, was renamed the Royal Dublin Golf Club in 1891 at the sanction of Queen Victoria.[10] By then, having first tried the coastal village of Sutton, it had moved to Bull Island in 1889, where it developed a links course in imitation of what existed in Scotland; this was a process of acquiring land around beaches that was repeated across the world.

Who owned this new island where Royal Dublin sought to build their club? Ownership of the island was a matter of dispute almost from the beginning. When the entity which later became known as the Dublin

Port and Docks Board (and later again, simply as Dublin Port) sought to build the North Bull Wall, it had acquired land from the local landlord, the Vernon family, who had been granted the foreshore under a charter from King Charles II in the seventeenth century. Following the construction of the North Bull Wall, the Port and Docks Board owned the wall itself (which was wide enough to promenade on) and a narrow stretch of land east of the wall. It was from these gathering sands that the island began to stretch out into the sea, growing with every season.

The ownership of this new land, named Bull Island, caused repeated controversy. As Colm Lennon wrote, "A popular opinion expressed in the press was that the island had been formed (accidentally) because of the expenditure of huge sums of public money on the North Bull Wall project and that therefore the area should be regarded as a commons, fully open to the public."[11] But, as the island ran parallel to the foreshore, claims on its ownership were made under law by local landlords: John Vernon and his family for the western part, and the Earl of Howth and his family for the remainder to the east. These claims on ownership ended up in court in the nineteenth century, before Lord Ardilaun, who lived at St. Annes, eventually bought out both the Vernons and the Earl of Howth between 1894 and 1902. Lord Ardilaun owed his enormous wealth to his inheritance of the Guinness brewery; he was also a philanthropist of considerable scale who had, among other things, bought, landscaped, and donated to the city the public park of St. Stephen's Green.

He now stipulated that Bull Island be a place of popular amenity; it was determined that the beach and its dunes should be open to public access. There was one exception: it was allowed that the Royal Dublin Golf Club should keep its course in a deal that amounted to ownership. Later, a second golf links was leased to the new St. Anne's Golf Club in the 1920s. These two private golf clubs extended the land that they owned or leased in a series of acquisitions which continued for more than a century. At its most crude, it was manifest in the installation of fencing which marked out their territory as private property.

There remained significant unanswered questions as to precisely how some decisions were taken that shifted so much land into the ownership or control of private clubs, but what is revealed time and again is the power of the golfers to secure title and expand their facilities. The Royal

Dublin Golf Club initially held a lease from the Vernon family, then held a ninety-nine-year lease of the grounds from Lord Ardilaun, as well as a portion in fee simple. This changed in the 1950s when a transaction allowed the club to purchase the island for £8,000. Royal Dublin duly retained ownership of their eighteen-hole course, as well as another piece of land to allow them to expand. Almost immediately, the club sold the remainder of the island to Dublin City Council for £12,795 in 1956.[12]

In the 1960s, St. Anne's Golf Club also took additional territory for its club. It was a nine-hole course at this point and the club wished to extend to a full eighteen holes.[13] The club was ordered by Dublin City Council to withdraw to its previous boundaries, but then in 1965, it secured an additional lease of what was said to be forty acres from Dublin City Council. In fact, the land to be given amounted to fifty-five acres and the decision to give it to the golf club brought a huge public outcry.[14] The expansion of the course was postponed, but eventually, by the 1980s, the golfers had secured enough additional land to build an eighteen-hole course. By the end of the twentieth century, the Royal Dublin Golf Club and the St. Anne's Golf Club essentially owned or held in long leasehold more than one-quarter of the land of the island (figure 14.1).

During these years of golfing expansion, there was tension between the competing desires of those who wished to extend the operations of the golf clubs of which they were members and those who saw the island as a popular site of recreation or as a nature reserve. It was not just that the presence of two golf clubs restricted wider access to a large section of the island, it was also the manner in which they conducted their activities. First, the initial operation of the golf clubs was broadly sympathetic to the island as a natural habitat, although it did immediately involve the removal of dunes and the leveling of other ground in the case of the Royal Dublin Golf Club.[15] As the twentieth century progressed, however, and the design, delivery, and maintenance of golf courses evolved, this meant the need to modify their space into an area that was artificially maintained and at odds with the nature reserve around it.[16] That Royal Dublin Golf Club became one of the most famous links courses in the world and that it hosted major international championships which drew the best golfers in the world placed on it a particular pressure to adopt methods using fertilizers and chemicals. This matter was set out by An Tasice, the Irish national heritage trust, in a report published in

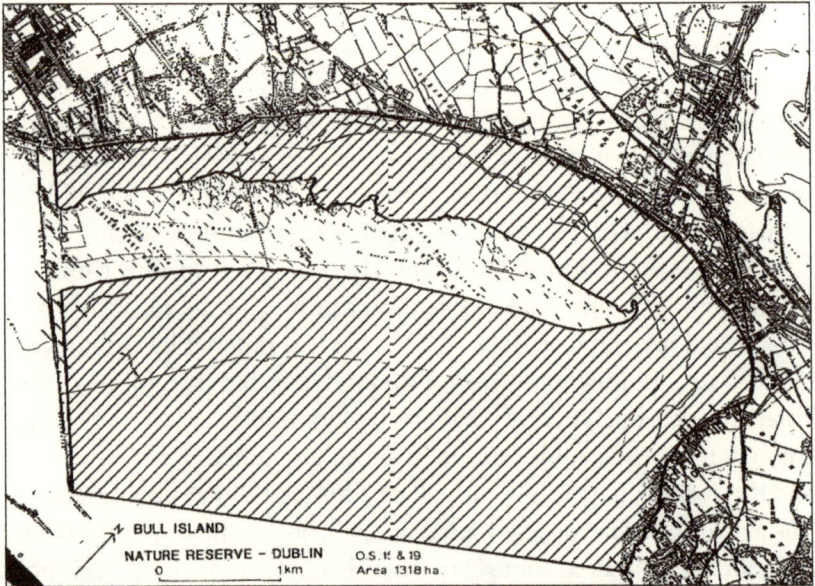

Figure 14.1. This map of Bull Island was adapted from a version found in the records of Dublin City Council, 1988. Courtesy of Dublin City Library and Archives.

1967 in response to ongoing attempts to expand the land on the island used for golf: "Further use for golf would cause serious damage to scientific values. . . . Any reclamation of slobland for golf would reduce the feeding area for birds. In addition, expansion of the golf course would damage the Island's considerable botanical wealth by the processes of seeding, use of selective weedkillers and all those practices necessary to management of golf courses."[17] There were further issues when, in the 1980s, Royal Dublin hosted the Irish Open golf championship, drawing leading professional golfers from around the world. Hosting the championship was made possible by extensive use of the beach as a carpark, the construction of a tented village for refreshment stalls, and the decision to locate a practice area on "the least disturbed, truly natural habitat south of the Bull Island causeway." Dr. D. W. Jeffrey of the Trinity College Dublin Environmental Sciences Unit wrote, "In my opinion this is virtually the worst possible site for the proposal."[18] Attempts to resolve the matter by the tournament organizers, local representative associations, the Dublin City Council, and four major conservation and wildlife groups demonstrated the extent to which holding a major modern sporting event on the island was inimical to its other aspects. After three stagings (1983–85), the competition was moved to a different venue.

Second, there were attempts by the golf clubs to privilege the interests of their members over the public good. This can be seen not just in the extension of their grounds as private property but also, for example, in how Royal Dublin Golf Club reconfigured its entrance road during the rebuilding of its clubhouse in 1954.[19] Later, in the 1970s, St. Anne's Golf Club developed a road to its clubhouse in a way that was "in serious breach of planning permission." That same clubhouse was also a matter of dispute and, although it ultimately received planning permission, it did so over the objections of heritage groups and after the failure to respond to letters from the Dublin City Council.[20]

The manner in which the golf clubs acted sometimes provoked a furious response from other visitors to the island and its strand. Indeed, in the mid-1950s, in response to yet more proposals to extend the golf courses, the Bull Island Amenities Preservation Society was founded and the desire was expressed to "prevent 300 golfers taking over what belongs to half a million Dubliners." There were suggestions that if Dubliners were fenced off from the beach by the golf clubs, "there will be bitter opposition. Fences may be torn down."[21] Ultimately, the issue was not that Dubliners were fenced off from the beach but that fences were erected by private golf clubs, which laid bare the extent to which they controlled access to a significant (and growing) area of the island and its dunes. Further, it revealed the manner in which local and national institutions of government did not have the vision, interest, capacity, resources, or legal tools to develop a plan for Bull Island that could reconcile its multiple uses.

Environmentalism, Sport, and Popular Recreation

Throughout the twentieth century, growing awareness of ecological and environmental imperatives revised the idea of what constituted acceptable practices. This eventually had a profound impact on the island and beach. It is a process that casts into the light not only how popular behaviors can change, but also how slow such change can be and how it can be resisted or ignored.

In the early 1900s, just as they had for decades previously, men came to Bull Island to shoot birds.[22] Those who shot on the island included the Lord Lieutenant of Ireland and other members of the elite of colonial Dublin society. They shot pigeons, snipe, pheasants, ducks, gulls, woodcock, herons, and all manner of other sea birds from cormorants to glossy ibises. Sometimes these birds were shot to allow them to be stuffed and put on display in museums or private homes; mostly, they were shot for sport. As evidence of the changing mores of society, there had been denunciation at the end of the nineteenth century of this "crying evil" of constant Sunday shooting on the island, but it was a practice that continued into the twentieth century.[23]

Activists and enthusiasts such as the Dublin Naturalists' Field Club (1885) and the Irish Society for the Protection of Birds (1904) argued repeatedly for a revised approach to the treatment of birds in Ireland. This led to the passage of the Wild Birds Protection Act in 1930 and then, in 1931, to Bull Island being declared as Ireland's first bird sanctuary.[24] The island was made a UNESCO biosphere reserve in 1981 and then a national nature reserve (with the exception of the land of both golf clubs, which was excluded from the order) in 1988.[25] In 1994, the island was declared an Area of Special Amenity by order of the Dublin

City Council, meaning that no development of any significance could be undertaken on any part of the island without permission.

Inevitably, this was no straightforward path to progressive policy; challenges came in various forms. First, a scheme to establish a marine lake and waterpark in the lagoon between the island and the mainland was proposed in the middle of the twentieth century. Initially, it centered on the creation of a motorboat racing circuit of three miles within an artificially created marine lake using a dam and other constructions that would develop the island as a commercial facility. This proposal was supported by numerous rowing clubs, yachting clubs, the Irish Swimming Association, the North Dublin Motor Boat Club, and the head of the Olympic Council of Ireland. These clubs endorsed a motion that declared it a matter of "urgent national importance" to develop an aquatic sports facility and "pleasure resort" on a site described as "an unsightly slob."[26] Through the 1930s, proposals were advanced—by making appeals to ideas of "national pride," "progress," and the imperative of creating employment, which chimed perfectly with the ideologies of the day—to build a pleasure ground in the manner of Coney Island and also a municipal aerodrome. After stalling with the advent of war, provisional approval was granted to proceed with the pleasure ground aspect of the scheme in 1946, on a site of eighty-six acres, to include an amusement park and a dance hall.[27] Opponents of the scheme condemned it as "vandalism" and fought valiantly to prevent it. This opposition, combined with the practical limitations of cost and the engineering challenges involved ultimately led to the plan being shelved after a tortuous campaign.[28]

Second, in the 1970s, having already overseen an enduringly controversial construction of a causeway from the mainland to the middle of the island, the Dublin City Council sought to respond to the increasing refuse disposal challenges it faced in the city by using "three acres of swampy ground on the bird sanctuary" as a dump. The plan was to fill rubbish into the area and then cover it with topsoil. That the land to be so "reclaimed" was under consideration to be given to St. Anne's Golf Club was also mooted.[29] The proposal angered environmental groups and some city councillors who were vehemently opposed. The council dumped refuse on the salt marsh until mid-1975, before stopping the practice.[30]

Third, old traditions did not just collapse in the face of change. At least until the 1970s, young greyhounds were blooded on Dollymount Strand by the pursuit of the (now extinct) hares that populated the dunes.[31] Also in that decade, five men and a woman were convicted and fined for illegally setting nets to catch hares. The nets were sixty feet wide, three feet high, and were set in the dunes. They had dogs and sticks and they put the caught hares into wooden boxes for transportation to coursing meetings around Ireland. The incidents were photographed by anti–blood sports activists.[32] There were also instances of hares being hunted by rifle, shotgun, crossbow, longbow, and slingshot along with their persistent mauling by dogs.[33]

Fourth, the wider public continued to engage in behaviors that undermined the island. For more than one hundred years, the Dublin City Council had repeatedly discussed the littering and vandalism of Dollymount Strand. By the end of the 1960s, an estimated ten thousand visitors came on long weekends. They left behind much of their rubbish. There were sanitary towels and condoms, empty bottles of beer, wine, and spirits, and all manner of discarded food wrappers. Children used glass bottles for impromptu cock-shots and left broken glass over the island. The aftermath of barbecues and other parties scarred the dunes. The beach became a carpark as the explosion of car ownership led more and more to drive out onto the sands; it was also the place where Dubliners learned to drive. Combined with the enduringly high levels of waste material in the sea water, it was enough to earn for the beach the name "Dirty Dollier."[34]

Nonetheless, that there was a shift in understanding of Bull Island was made clear in the reaction to various developmental proposals. When the 1970s saw proposals for a third golf course, a pitch-and-putt course, and football pitches, there was criticism that the last sense of its wilderness would be destroyed, and it was argued that sport should not be allowed to destroy a place that "unexpectedly came from the sea to the people over the last 200 years; it should be protected and controlled for the sensible recreation of Dublin people and as a haven for wildlife." The proposals were defeated.[35]

A New Millennium

Bull Island and Dollymount Strand demonstrate how the accretion of change and the deepening of traditions are what make the present. When you turn right off the Coast Road, coming out from Dublin city, and cross the wooden bridge onto the North Bull Wall, you pass down along the front of Royal Dublin Golf Club. There is a barrier across the entrance to its carpark and a large sign reads "Private Property." In December 2021, the Royal Dublin Golf Club voted to stop being the last remaining male-only golf club in Ireland when its members voted to accept new members from "all genders." Although there had previously been no written rule forbidding female membership and there had been female golfing competitions staged, a ban on female membership had existed in practice.[36] Further down the island, the Dublin City Council continues to engage with St. Anne's Golf Club around the precise nature of its boundary. Despite the dramatically increased awareness among golfing institutions of their courses as environmental habitats and the consequent changes in behavior in respect to course management, the operation of the golf clubs remains a concern to the biosphere.[37] For example, the extraction of fresh groundwater for irrigation purposes has a potentially serious impact on hydrology, which plays an important role in the island's ecology, and the location of the St Anne's Golf Club wastewater treatment near the marshes on Bull Island is recognized as a potential hazard.[38]

With the exception of the golf clubs and the Bull Wall (with its modernist concrete bathing shelters, built in the 1930s), the rest of the island is now owned by the Dublin City Council and is the largest public park in the city.[39] The bathing shelters are a legacy of the commitment that the Dublin City Council had to providing simple public amenities. The capacity of the city council was limited by the resources at its disposal and by the prioritization of other needs, but these bathing shelters stand as monuments to the provision of public amenities.

Some of the people who swim from the concrete shelters also compete in open-sea swimming competitions and in triathlons for adults and children that are now run along Dollymount Strand. The strand also hosts fun runs, 5k races, and water festivals with beach volleyball, paddleboard racing, and other aquatic sports. It continues to be a place of sporting innovation. For example, kite surfers now make use of the beach to shoot up and down the shallow water at speeds touching sixty kmph. Kite surfing has become one of the fastest growing water sports in Ireland, whose participants organize themselves as a community rather than into a club. On Dollymount, there are Irish men and women who have taken up the sport in recent years, and all around them are Poles and Lithuanians, Germans and South Americans, immigrants who work in a city thriving around the American technology companies housed in its docklands.

Most of all, though, the strand is a place for popular recreation. By 2020, there were an estimated 1.4 million annual visits to the island. What historian David Dickson denoted as "the revolutionary cheapening of distance" created a new context for trips to Dollymount Strand; package tourism, sun holidays, and the emergence of cheap air travel from Ireland in the 1990s transformed the Irish understanding of trips to the seaside.[40] Recreation along the coastline of Dublin Bay has been redrawn as the importance of family outings to the seaside has declined; the horizon of "the beach" has been extended far beyond Dollymount, not least because of the dramatically increased wealth in the country. But Dublin Bay remains fundamental to recreation and is by a considerable distance the "largest public amenity on the east coast of Ireland," while Dollymount Strand is its most popular beach.[41]

The island has now been awarded the greatest number of conservation designations of any site in Ireland. It has areas and species that are protected under the EU Habitats Directive. It is internationally acknowledged as a key part of the flyway for migrant birds traveling from the Canadian Arctic to the Mediterranean region and Africa. Its wildfowl and wading birds are protected under the EU Birds Directive, and it also supports birds such as sand martins, swifts, and swallows, which migrate to breed in Ireland. The island's flora contains rare and legally protected plant species identified in a landmark Dublin Naturalists' Field Club report from 2019 (figure 14.2).

This is a complicated story, however. Broader societal awareness of environmental issues has never been higher and yet "a significant percentage of the people exercising on the island, either with or without dogs, do so without the realization of the sensitive dynamics of the ecosystem in which they are recreating. These are not irresponsible people.

Figure 14.2. Dating from the mid-1950s, this aerial photograph of Bull Island was taken by the pilot Captain Alexander Campbell "Monkey" Morgan. It provides a bird's eye view of the full stretch of Dollymount Strand and the golf courses behind it on Bull Island. At the bottom, is the North Bull Wall, to the right is Dublin Bay and to the left is the lagoon which separates the island from Dublin city. Courtesy of the National Library of Ireland, Morgan Aerial Photographs Collection.

They just do not have a full appreciation and understanding of the sensitivities of this particular nature reserve."[42] Accordingly, the island's hare population is now extinct, its seal population is routinely disturbed, and its ground-nesting birds have their habitats destroyed; it has not escaped the national and international collapse in biodiversity. That there remains a need for regular beach cleaning evidences the enduring capacity of people to litter.

Ultimately, the greater context in which the history of the beach, its pleasure seekers, its sportspeople, and its nature lovers will be forged is the wider story of climate change. The most recent surveys from scientists project that by 2050, ongoing environmental damage will see Dublin Bay's coastal area suffer extensive flooding. The rise in the global mean sea level as set out in the 2019 *Special Report on the Ocean and Cryosphere in a Changing Climate* by the Intergovernmental Panel on Climate Change (IPCC) has profound implications. The projected depths of inundation render the Strand and Bull Island hugely vulnerable. The place that grew from the sea in the wake of human intervention may soon revert to from whence it came through human intervention of a more destructive type.

NOTES

1   The writer Brendan Behan described Dollymount as Dublin's Riviera: Andrew McNeillie and Brendan Behan, "The Dublin End: Anecdotes of Brendan Behan on Árainn," *Irish University Review*, Special issue, *Brendan Behan* 44, no. 1, (Spring/Summer 2014): 72. I would like to thank Shane Derby, the staff of Dublin City Council and Fulbright Ireland for their assistance with this chapter.

2   "Sporting Intelligence," *Freeman's Journal*, June 11, 1850, 2.

3   "The International Rifle Match," *Irish Times*, June 16, 1875, 3.

4   "After the Battle," *Irish Times*, June 30, 1875, 5.

5   John Rigby, "The Late International Rifle Match," *Irish Times*, July 17, 1875, 5.

6   "The Great International Rifle Match," *Irish Times*, June 30, 1875, 5.

7   David Burgess, "Tough Start to World Championship," *Irish Times*, July 20, 1977, 3.

8   George Burrows, "Off the Reel," *Irish Times*, May 16, 1964, 5.

9   Kieran McNally, *The Island Imagined by the Sea* (Dublin: Liffey Press, 2014), 172–73.

10   *Royal Dublin Golf Club Archives*, Letter from British Home Office to Royal Dublin Golf Club, June 12, 1962.

11   Colm Lennon, *That Field of Glory: The Story of Clontarf from Battleground to Garden Suburb* (Dublin: Wordwell, 2014), 194–95.

12   Dublin City Council minutes, Report of An Coisde Airgeadais, February, 27, 1956.

13   McNally, *Island*, 160–61; Lennon, *Clontarf*, 228–29.

14   See for example Dublin City Council Report of An Coisde Pleanála agus Forbairte, July 13, 1965 and minutes, June 13, 1966.

15   "Golf," *Irish Times*, Octtober 22, 1900.

16   *Royal Dublin Golf Club, 1885–1985: Diamond Jubilee Year* (Dublin: Royal Dublin, 1985), 18; *The Royal Dublin Golf Club, 1885–1985: Centenary Year* (Dublin: Royal Dublin, 1985), 46–55.

17   Anthony J. Wickham, "Bull Island," *Irish Times*, April 14, 1969, 11.

18   Dick Grogan, "Golf Practice Area at Bull Island Opposed," *Irish Times*, June 14, 1984, 7.

19   "Traditions of Royal Dublin Golf Club Are Well Maintained in New Clubhouse," *Irish Times*, October 4, 1954, 3.

20   McNally, *Island*, 162–63.

21   "Bridehead on the Beach—Plans Mr. Robbins," *Irish Times*, December 24, 1955, 3.

22   McNally, *Island*, 112.

23   McNally, *Island*, 87.

24   "The Protection of Birds: Great Progress in Ireland," *Irish Times*, May 11, 1932, 4; National Archives of Ireland, D/T 97/9/587 North Bull Bird Sanctuary Dollymount and D/T 2002/14/1279, Wild Birds (North Bull Island) Order.

25   National Archives of Ireland, D/T 2018/3/859, Nature Reserve (Bull Island) Establishment Order, 1988 and AGO 2018/4/99 Nature Reserve (Bull Island) Recognition Order, 1988.

26   "Marine Lake for Dublin: Aquatic Clubs' Proposal," *Irish Times*, August 29, 1930, 4.

27   "North Bull Island as Recreational Centre," *Irish Times*, November 7, 1946, 7.

28   McNally, *Island*, 146–47.

29   George Burrows, "Bull Island Dumping Plan Angers Groups," *Irish Times*, January 20, 1971, 9.

30   McNally, *Island*, 87.

31   Dublin City Council minutes, June 12, 1967; "Island Story," *Irish Times*, May 10, 1974, 13.

32   "Six Fined for Netting on Bull Island," *Irish Times*, February 16, 1971, 13.

33   Cóilín MacLochlainn, "Wild Hares Released on Bull Island," *Irish Times*, May 4, 1995, 2.

34   Dublin City Council minutes, January 7, 1907, April 13, 1953, and November 3, 1980; Elisabeth Leslie, "The Road to 'L,'" *Irish Times*, June 23, 1961, 8; [Quidnunc], "An Irishman's Diary," *Irish Times*, August 1, 1964, 9; "Youths for Trial on Rape Charge," *Irish Times*, July 19, 1967, 4; "Beaches to Have Litter Wardens," May 31, 1969, 11; Dick Ahlstrom, "Dollymount Water 'Near-Perfect,' Says Corporation," *Irish Times*, June 1, 1989, 5; "Bathing Areas Criticised," *Irish Times*, August 4, 1990, 2; Lara Macmillan, "Dirty Dollier Shocks City Councillors," *Evening Herald*, September 24, 1992, 8.

35   "Island Story," *Irish Times*, May 10, 1974, 13.

36   Brian Keogh, "Royal Dublin Votes to Accept Women Members," *Irish Independent*, December 2, 2021, 34.

37   *Environmental Policy Statement of Royal Dublin Golf Club* (Dublin: n.p., 1999), 1–27.

38   Parks, Biodiversity and Landscape Services Dublin City Council, *North Bull Island Nature Reserve: Action Plan 2020–2025 for the Implementation of Management Objectives* (Dublin: Dublin City Council, 2020).

39   Paul Melia, "Bull Island land to be transferred into public ownership," *Irish Independent*, September 18, 2013, 5.

40   David Dickson, *Dublin: The Making of a Capital City* (London: Profile, 2015), 543.

41   Richard Nairn, David Jeffrey, and Rob Goodbody, *Dublin Bay: Nature and History* (Cork: Collins Press), 1.

42   Parks, *Action Plan 2020–2025*, 12.

15

Regulation and Freedom on Europe's "City Beaches"

QUENTIN STEVENS

A "natural" beach is, in principle, part of nature's bounty, a collective, public good until made otherwise, and the politics of most beaches revolve around how to manage them, how to fairly allocate their benefits, and how to cover the costs of providing them. Many beaches are shaped by human action. But the beaches discussed in this chapter are completely artificial, a product of work by local governments, community groups, and private businesses. Natural beaches usually provide public views and access to oceans and rivers. But these cases are hundreds of kilometers from the coast and only some provide direct access into water, usually by constructing pools.

The characteristics of these artificial city beaches vary, but their key physical attributes include an urban location, a large volume of sand spread on an open space, a view onto water, and the inclusion of objects thematically associated with beaches, including deck chairs, beach umbrellas, palm trees, and thatched huts. They create the sensory and social atmosphere of real beaches within urban settings that would otherwise lack them. The types of sites transformed into these artificial beaches include cleared industrial land, rail easements, parking lots, roofs of parking garages, public plazas, and low-use sections of parks. By utilizing such sites, city beaches supplement the scope of urban open space. These beaches attract a broad demographic for informal socializing, playing, engaging in sports, attending programmed cultural events, and casually drinking and dining. Two other defining characteristics of artificial beaches are their temporariness and mobility. Most only operate in the warmer months and are removed and stored during the winter.

This chapter examines the distinctive politics of a range of artificial beaches in Germany and France, two large Northern European countries with large inland populations. German cities have hundreds of artificial beaches, mostly installed and managed by private entrepreneurs as hospitality venues, with others produced by community groups or artists. In France, most artificial beaches are created by local governments to serve local residents.

The politics of these beaches are distinctive because of their different relations to urban spatial contexts and the distinctive public and private inputs into their creation, ownership, management, and use. These beaches are often very accessible, situated right in the hearts of cities and neighborhoods, on or adjacent to the liveliest parts of the public realm: downtown plazas, parks, shopping streets, and malls. Some are on private property and have fences; a kind of privately owned, publicly accessible space. Their amenity and publicness vary, influenced by both legislation and management practices. Others are public ventures on public land, and are created to increase amenity, social engagement, and escapism for all. The management of some artificial beaches is very participatory and democratic. Not all artificial beaches are created and reserved for wealthy residents and tourists.

Artificial beaches are not privatized real beaches. They are rarely the subject of large-scale, high-profile public protests or legal disputes. But many of them represent the channeling of economic, legal, and political means to regulate and privatize urban spaces, their management, and economic flows so that some individuals can capture wealth and political power. Artificial beaches also raise questions about restrictions of access and use, citizen's rights, the just allocation of costs and benefits, and the atmosphere of freedom that most beach environments signify. By blurring traditional distinctions of public and private spaces, behaviors, and rights, artificial beaches also potentially create new opportunities for real public life and community identity. City beaches can, as the Situationists imagined, also contribute to the realization of the fundamental right to the city.[1]

France's Public City Beaches: *Sous les pavés, la plage!*

The idea of creating a beach within the city has a revolutionary origin in the calls of the Situationist International during the left-wing Paris uprising of May 1968 to unearth "the beach under the paving stones."[2] For the Situationists, this was a political vision. The loose, sandy substrate underneath the functional, hard-surfaced pavement symbolized the possibility of citizens reclaiming and reimagining the city, its social life, and the definition of leisure.

The hundreds of artificial city beaches that have been erected in Europe and worldwide in recent decades owe their inspiration to a single project in small, provincial St. Quentin, on the Somme River in France's northernmost and most economically disadvantaged region. The city's assistant mayor for animation and development, Xavier Bertrand, noted that its large town square stood deserted and that few residents could afford summer vacations (the coast is 150 kilometers away). In summer 1996, Bertrand transformed the square with sand dunes, pools, palm trees, deck chairs, play equipment, and entertainment so that local families could enjoy a beach experience locally. It was hugely popular and has continued annually, later expanding to animation of a second, permanent beach site, by the Somme near the railway station. After each two-month summer deployment, St. Quentin lends their portable beach infrastructure to Hirson, 80 kilometers west.[3]

In 2001, Socialist Bertrand Delanoë was elected mayor of Paris, having campaigned to reduce automobile use and make the city more livable for residents. After he temporarily trialed closing the Pompidou Expressway alongside the Seine for one month in summer 2001, in 2002, he appointed set designer Jean-Christophe Choblet to develop a beach-themed transformation of that riverside space, amid Paris's key world heritage attractions around the Île de la Cité. Two-thirds of its original €2 million budget was covered by sponsors. As in St. Quentin, this beach's aim was to enhance social inclusivity and engagement, "give the riverside back to Parisians," and provide "a nice hangout [where] people, with their differences, will mingle," especially poor suburbanites "who never leave [Paris] on vacation."[4] *Paris Plage* attracted two million visitors during its first month. It later expanded to several other sites (figure 15.1): beach volleyball outside the city hall in 2003; the Joséphine Baker Pool floating in the Seine outside the Bibliothèque François Mitterrand in 2006; and the Bassin de la Villette, an eight-hundred-meter-long widened section of a northern Paris canal, later embellished with three floating pools in 2007. The Seine pool uses treated river water; the canal's water is adequately clean. Although *Paris Plage* became famous and is a significant tourist drawcard, well over half its visitors are Parisians.[5]

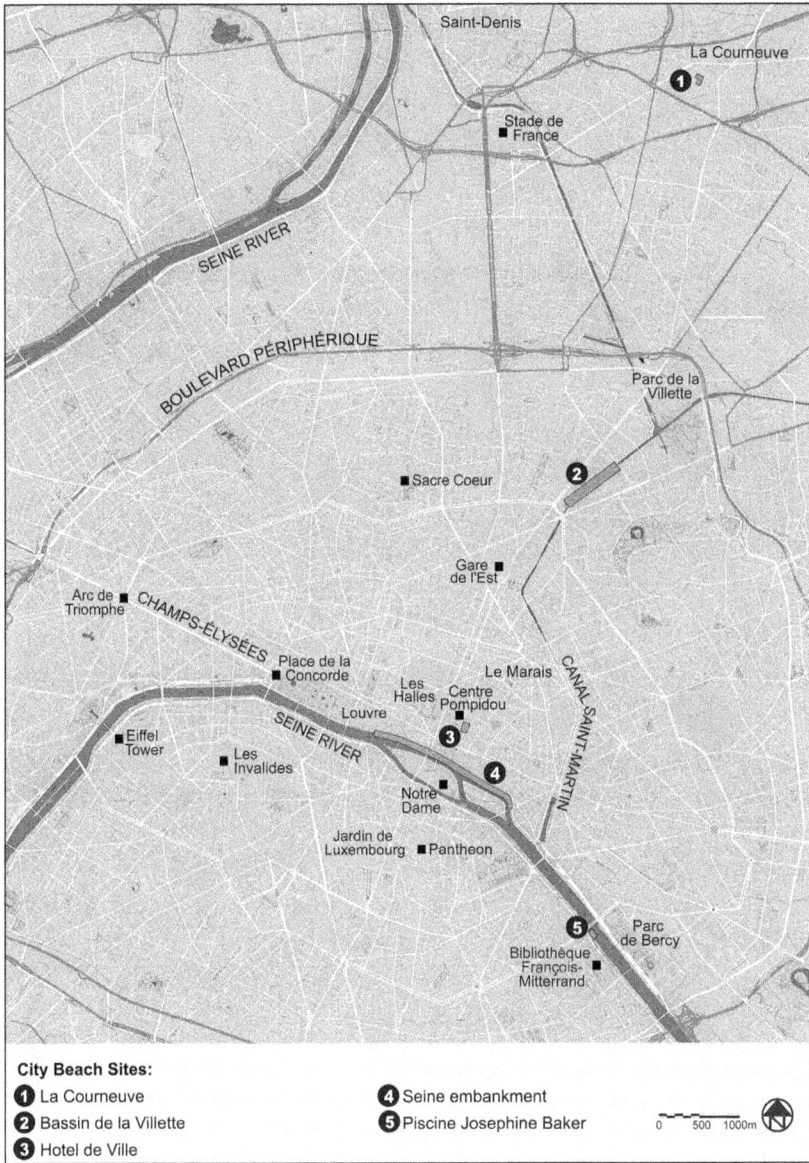

Figure 15.1. Map of city beaches in Paris. Courtesy of Quentin Stevens and Mohammad Mohammadi.

Figure 15.2. City beach in La Courneuve, suburban Paris. Photo by Quentin Stevens.

These beach projects are tangible examples of French *egalité*, providing a beach experience to those residents least able to enjoy the real thing. This is the case for both St. Quentin, governed by the center-right, and Paris, governed by Socialists. By summer 2005, local governments had established artificial beaches in many provincial French cities, towns, and poor suburban Parisian *banlieue*, including amid the high-rise apartments of deprived, crime-ridden, largely African and Arab La Courneuve (figure 15.2).[6] With resources limited, La Courneuve's deputy mayor asked the local fire brigade to fill the swimming pool and convinced construction companies to donate sand.[7]

Grassroots Beaches in Germany

The politics of *egalité* also motivated Germany's first-ever artificial city beach in Vaihingen an der Enz, near Stuttgart. In 2001, this small city's council took inspiration from an earlier government-funded workshop involving architects and artists, which had proposed transforming the town square with a sandpit beach. Since 2003, a local citizen initiative has installed and managed *Strandleben* (beach life) every second summer using donated materials and labor. The project's key aims have always been to playfully reimagine the social potential of public space and to increase local citizens' active engagement. The organizers emphasize that *Strandleben* should both serve local residents and help local businesses: "Our association organizes this project for the children, for the mothers and fathers during the holidays . . . [and] also . . . so the city center gets more publicity outside of Vaihingen. So that we get visitors . . . to stimulate the city center, in terms of the people [and] the economy. But the condition is that it's public, nothing exclusive where one has to pay admission fees."[8] The organizers also emphasize the importance of residents actively engaging in creating and managing their city beach:

> What was good was that a lot of people were involved. . . . We have fliers: "What can I do for the Vaihingen beach?" There's donations and accepting cleaning shifts and organizing events, et cetera. We told [residents] that if they want to have it, they will have to become active and do certain things. I think the good thing is that the people of Vaihingen consider this to be *their* beach. They look out for each other and whether everything is running okay. I think that this is the secret to success. No one simply set it up. One had to fight for it and one had to do something for this beach every day. They identify with this beach.[9]

In nearby Esslingen, another community-based beach was established in 2009 by a group of youth work associations. They felt local adolescents lacked outdoor activity options and wanted to motivate youths to exercise. Their beach was installed rent-free on city-owned riverside parkland and was supported by a community fund and sponsorships. It was supervised by paid and volunteer educators and frequently offered programs with participatory activities. When the temporary beach was repeated in 2010, adolescents who had not succeeded in obtaining an apprenticeship were trained and supervised to rebuild and decorate a Christmas market stall as a beach cabin, which gave them a sense of pride and encouraged them to visit the beach regularly.[10]

In 2011, the official opening of a new neighborhood park with climbing walls in Schwedt was accompanied by a temporary beach club with artificial palms and nonalcoholic drinks served by youth from the regional youth development association. This project arose from a discussion between a local youth club and a visiting state politician that highlighted the shortage of meeting places for youth. Within four weeks, the youth themselves had planned the project, acquired sponsorship, and installed the beach. The politician donated some deck chairs.[11]

A fourth socially motivated beach project, Munich's *Kulturstrand*, was created by the Urbanauten, a group of university students, following discussions about how to encourage residents to see and use public spaces differently and how to modify leftover spaces into venues for performances, cultural events, public debates, and social engagement. In 2005, they applied for planning permission for a temporary public beach where people could bring their own refreshments, and they unexpectedly received a one-month permit. Lacking any funding, they contacted a restaurateur they knew, who provided financing and operated a bar on the beach. The Urbanauten involve local organizations and reinvest much of their financial turnover into the beach's daily programming of live music, film, theater, and children's activities. The Urbanauten have subsequently relocated their beach and activated six different locations. *Kulturstrand*'s contribution to local social life is recognized by Munich's mayor, their official patron. It has brought new ideas and activities to little-used parts of Munich's riverfront and encouraged participation and debate about urban space and public life.[12]

The emphasis of these grassroots city beaches is social engagement. While some lack water access or even sand, potted palms and beach furniture adequately evoke the desired relaxed atmosphere.[13]

Privately Owned "City Beaches" in Germany

Only a few city beaches in Germany have been developed for local residents by public or nonprofit agencies. Hundreds more have been created by entrepreneurs from the hospitality, entertainment, and media industries (figure 15.3). *Strandbar Mitte*, Germany's first commercial city beach, opened in Berlin in 2002 just one month after *Paris Plage*. Like *Paris Plage*, it sits amid world heritage museums and palaces, on a little-used space alongside the Spree River, Monbijou Park. It was created by a theater manager who had run an open-air stage in the park since 1999. He wanted to sell food and drink to patrons, rather than losing this

Figure 15.3. *Bundespressestrand* ("Federal Press beach"), a commercially run city beach that, from 2003 to 2011, occupied vacant land in Berlin's government quarter subsequently developed for new government offices. Signs to patrons note, "Dear guests, please bring your good mood"; "Leave your food and drinks at home!"; "Please don't walk barefoot!"; and "Splinter warning!" Photomontage by Quentin Stevens.

potential income to nearby hospitality venues. After seeing *Paris Plage*, he obtained a liquor license, brought seventy tons of sand from a lake outside Berlin, and opened an outdoor bar with deck chairs and potted palm trees. *Strandbar Mitte* was so profitable that the next summer, the owner established a much bigger beach bar, *Oststrand*, on the empty strip of riverfront land behind the last remaining stretch of the former Berlin Wall, the East Side Gallery, decorated with artists' murals.[14] This entrepreneur has enlivened little-used spaces in a city that had lacked public resources and created many hospitality jobs. Like *Paris Plage*, these beaches also enhanced central Berlin's image for tourists and locals and created a model for many similar initiatives elsewhere. In the following few years, Berlin alone gained over seventy similar privately run beaches. Most large German cities had multiple city beaches. Most of them sell beer and cocktails from thatched huts or shipping containers. Some are run by chain operators.

City beaches became particularly popular during Germany's hosting of the FIFA Football World Cup in 2006. This was a record warm summer and many beaches hosted public viewings of live matches, with large audiences seated on deck chairs watching outdoor screens. Some commercial beaches were installed on public parkland with water frontage. Many were on former industrial sites. Some, like *Strandbar Mitte*, were installed adjacent to existing bars or institutions such as theaters and universities. These beaches enlivened a diversity of marginal, underutilized urban sites, including floodplains, railway easements, and spaces underneath bridges. Numerous beaches were installed on the sunny rooftops of multistory shopping center carparks.[15] These private investments have created attractive, lively open spaces on formerly empty, privately owned land. Some local governments have offered free land and logistics support and streamlined approval processes to attract entrepreneurs to create and run city beaches to bring new attention and life and attract new investment to industrial brownfields. The politics of these vibrant pop-up beaches often align to neoliberal place marketing and gentrification. Other criticisms have included local governments temporary variances to local bylaws, excessive noise and garbage, bad behavior of departing patrons, and the appropriation of public sites for commercial enterprises.[16]

The nadir of privatization of artificial beaches is the entirely enclosed, domed, and heated artificial beach of Tropical Islands Resort, built in 2004 inside a massive former zeppelin hangar sixty kilometers south of Berlin by a Malaysian cruise ship magnate. Here, visitors pay by the hour to sunbathe, swim, and camp overnight inside a pleasurable environment kept at 25°C year-round. Polynesian dancers perform nightly during the artificial sunset and dinner show and must live for the winter in a small adjacent hotel. This is a postindustrial form of colonialism, where the tropical island, its inhabitants, and their culture have all been captured and reproduced in Northern Europe under glass.[17]

Politics in the Making

The politics of these varied artificial beach projects are complex and diverse. As environments whose form and image are entirely shaped by human action, we can critique the politics of creating artificial beaches as well as the politics of managing and using them.

Artificial beaches have been produced by diverse sets of actors, with varied motives and inspirations, including community groups, entrepreneurs from the hospitality and creative industries, and local governments.[18] Artificial beaches require substantial inputs of land, labor, and capital, and these are inevitably targeted toward maximizing benefits for particular user groups. The rights and freedoms of each artificial beach are allocated before and through its creation, defined through a series of contracts, purchases, regulations, and boundaries.

Community-driven city beaches involve close, specially negotiated citizen-city government relationships. Community-based facilitators of these projects, which include residents, youth, churches, and arts organizations, maintain open communication, engagement, and trust with local governments, who provide in-kind support and grants for space, equipment, and logistics as well as approving permits and adjusting or regulations. The community organizers of Vaihingen's beach highlighted that the relationships that drove it did not exist beforehand. Trust developed through the process of making the space, through detailed discussion and negotiation over rights and responsibilities: "We meet in advance and talk about everything. Support and responsibility are shared between the council, the community association, and organizers of individual events. . . . We talk [to event organizers] about the conditions, so that they are actually responsible for this themselves. . . . The *Ordnungsamt* [Public Order Office] trusts us with this."[19] The politics of rights and responsibilities on artificial city beaches are shaped by relationships and actions that extend well beyond their immediate time and space.[20] These include deindustrialized waterfront sites, local governments who want to loosen up rigid planning controls, and managers of indoor hospitality venues who lack business during the summer. Many spaces these beaches occupied were planned for redevelopment for upmarket hotels, apartments, offices, and leisure venues. This open space format was thus strongly influenced by the temporary time window between long-term urban development cycles.[21]

The lure of profits has encouraged many entrepreneurs to create spectacular, hedonistic artificial beach landscapes that appeal to wealthier urban residents' enthusiasm for open spaces, novelty, leisure activities, and display. In many cases, entrepreneurial operators of commercial city beaches extract significant profits by renting spaces and equipment cheaply, employing staff casually, circumventing expensive regulatory requirements, and externalizing costs onto local governments and communities. But not all city beach managers are wealthy businesspeople. City beaches are not necessarily profitable, and capital is often very limited. The lack of market demand and financing to redevelop Germany's extensive urban brownfields has created opportunities for new kinds of managers with very small budgets and time horizons. Some beach operators are "dropouts" seeking escape from a life driven by economic imperatives.[22] Landowners often pursue beaches as tenants, with low or no rents, to help advertise their site or to cover their holding costs. Market weakness also means that city beaches' production and operation, including their admissions policies, use programs, and employment, remain strongly guided by local governments. German city governments do not just regulate privately instigated beach proposals; many cities proactively explore, market, and underwrite these opportunities through grants, loans, rent-free land, free services, and tax reductions.[23]

The short-term, dynamic, sometimes haphazard arrangements that constitute city beaches, and their physical looseness, allow many opportunities for various actors to modify and extend these settings, rules, and uses. Rights and responsibilities to shape city beaches are dynamically created, assigned, and transferred. Beaches have been bought and sold between different private, public, and community operators and even relocated when their sites were redeveloped. Munich's *Kulturstrand* highlights the constantly negotiated politics of the city beach. Responding to resident concerns about noise, garbage, and traffic, its organizers negotiated with the city to rotate annually between four sites. Each local context influenced its spatial form, programming, and operational constraints.[24] These beaches can shift flexibly with economic and political tides.

The Management of Artificial Beaches

Medieval King Canute could not hold back the tide. But the managers of artificial beaches can exert near-total control over their social and environmental dynamics. Despite the strong formal similarities among city beaches, differences among the human actors that run them precipitate differences in the publics drawn to these spaces and the actions and social relationships that these spaces afford and allow. However, these differences are not as simple as saying publicly run city beaches are open and tolerant and commercial ones are controlling and exploitative. The extent of control varies and so does its aims.

On publicly funded and community-run beaches, numerous vigilant, uniformed animation staff and security guards ensure child safety and cleanliness, and there are often bans on smoking and alcohol. When I first viewed and photographed the community-run city beach in Vaihingen, Germany, I was approached by a resident who amiably asked whether I was visiting. Their role as a community association member was to keep the beach tidy and manage any problems. At every publicly run city beach I visited in France (three in Paris, two in St. Quentin), site managers approached to inquire who I was and why I was taking photographs. Some of them were concerned I was infringing on the privacy of individuals. Others asked whether I had official permission to take and publish photographs of the beach. These public beach managers want to control the narrative and protect the brand. In the case of St. Quentin, I was granted a research interview with the senior city council administrator who managed the beach and permission to photograph it on condition that I only had positive things to say about it. There is no doubt the beach in St. Quentin is good and popular. But what are the politics of being free to photograph it and to have an opinion about it? Public city beaches appear to be relatively paternalistic and risk averse.

By contrast, on Germany's many private, commercially run city beaches, there are few staff beyond waiters, and the main behavioral control is preventing patrons from bringing and consuming their own drinks (see figure 15.3). Germany's artificial beaches are generally very open environments where people can go and act as they wish. Pools are unsupervised. People can sit and swim for free. Visitors do not have to purchase drinks. Nevertheless, these beaches' core purpose is hospitality. Some commercial beaches, like some live music venues, have a *Minimumverzehr*—on entry, visitors must purchase a voucher valid for food and drinks. The second main management control is that visitors cannot impair other visitors' enjoyment of the beach. This involves (very rarely) removing unruly customers. Signs note that management takes no responsibility for customers' belongings, but they do watch out that customers are not destroying the atmosphere. During the hundreds of visits I made to hospitality-run city beaches, I was only once asked by a manager why I was taking so many photographs, and they were actually glad about the free publicity.

The Five Freedoms of the Artificial Beach

To understand the politics of *using* artificial beaches—how open, public, and "free" they are—it is useful to consider Kevin Lynch's articulation of the five kinds of rights that define control over spaces and whether these rights are fairly allocated. As a key theorist and critic of how well urban environments address various human values and needs, Lynch argues that user control and the adaptability of environments are neglected values within city planning and that equity of access and control in urban spaces are among "the most crucial areas of environmental justice."[25] The socially constructed character of artificial beaches provides opportunities to debate and adjust these freedoms.

The Right of Presence

Artificial city beaches mostly maintain the right of presence. Even when privately built, managed, and fenced in, they generally have free admission and are just as open to all users as public beaches. Local government approvals often require they remain freely accessible. Even on commercial city beaches, visitors can sit in the sun and enjoy the atmosphere for free. Pragmatically, most of these projects seek to welcome a diverse demographic to help cover their costs. Only on rare occasions must visitors pay admission, for example, for live performances or on weekend evenings when owners believe they can profit most from long-staying, big-spending groups of patrons. I have only once seen an unruly patron ejected from a city beach. That does, however, prove the rule that private management includes the right to refuse admission. A more significant exclusion is that artificial beaches are only accessible when their managers open their gates and provide casual staff—when they estimate they will generate income from hospitality sales. Their fixed costs are very low. In practice, this means many city beaches only open on sunny afternoons and warm summer evenings, sometimes only on weekends. Visitors are excluded from experiencing city beaches alone on cold, rainy mornings.

The right to presence also depends on the attitudes of their operators and users. When 49 commercial city beach managers were asked whether their project served the ambition of "creating a public space," only 75 percent agreed, and very few articulated this social ideal in detail. One manager explained that their commercial city beach was "definitely a public space" because "it belongs to the [public] park that belongs to everyone . . . everyone can come here no matter whether they buy something here or not. This is still a public space that we can't simply restrict . . . we're not a mere money machine. One has to find a compromise."[26] Insights about city beach users' perceptions of their publicness can be drawn from two large-scale surveys undertaken in Germany. Stefanie Kahls asked 298 visitors to four German city beaches why they preferred visiting a city beach rather than a conventional bar or beer garden.[27] Only 16 percent of visitors valued the specific public that city beaches attract, and only 7 percent valued the ease of making social contacts. Almost all visitors came with friends, colleagues, or family members and spent extensive time there, and on one particular small-city beach, in Trier, over 80 percent of these visitors brought their children. That beach's child-friendliness was rated one of its top four advantages over bars. In this sense, beaches are more inclusive than many privately run, indoor "third places."[28] Nevertheless, most people on the four beaches Kahls studied were locals who came to relax with people they knew.

Sachs's survey of 827 visitors to six city beaches similarly found that half came with friends and a quarter with their partners. Most interviewees (90 percent) felt city beaches were relaxed and *gemütlich* ("comfy," "friendly"). Sachs characterizes the city beach as "a (semi-) public space . . . (where) one can be 'easy going,' one wants to, and can, relax and switch off—a quality that is apparently not available in conventional hospitality venues to the desired degree."[29] Even commercialized beaches thus appear to provide a distinctive, inclusive social atmosphere.

The Right to Use and Action

Controls on visitors' behavior within city beaches are minimal, although some subareas are defined for particular uses, such as dining tables, dance floors, and beach volleyball courts. One key control is the implied contract, articulated by signs at most city beaches' entry gates, that visitors cannot consume drinks they have brought with them. At some beaches, staff check bags and make people leave water bottles outside. This constraint, and the use of minimum-purchase vouchers, reflect the fact that income from food and drink sales pay for the beaches' very existence. The deck chairs and umbrellas are usually provided free by sponsoring drink and cigarette companies. These beaches are only loosely user-pays; Tropical Islands Resort is an exception. These beaches are certainly different to conventional sidewalk café tables that are customer-only. Toilets at city beaches are usually staffed and require payment, but this is typical throughout Germany. The second key regulation of use is that visitors cannot disrupt the enjoyment of other visitors. Again, "enjoyment" here should be understood as the consumption of drinks and food. On balance, these artificial beaches leverage private investment and consumption spending to provide accessible landscape amenities for broader publics.

The Right of Appropriation

Appropriation involves users' ability to define and monopolize territories for specific users and activities and exclude others from them. An artificial beach is itself a fenced-in appropriation of urban space. Some beaches do monopolize sections of inner-city riverfronts. But in most cases their sites were previously not publicly accessible, or they only temporarily transform a little-used portion of a larger public park or plaza. This appropriation of underused spaces is often not zero-sum; it creates new spaces rather than closing off existing ones. Some beaches, applying a "freemium" model, include luxurious, rentable day beds and "VIP" areas for private parties. Public beaches, by contrast, cannot define some people as more important. Sometimes artificial beaches are rearranged by management for particular larger scale events, especially for public viewing of major soccer matches on large screens from rows of deck chairs on summer evenings. This does inhibit other uses, but not more so or more often than similar public events that temporarily appropriate public parks.

The Right to Modification

The right to modification means people can physically change a place and prevent others from doing so. In this context, artificial beaches are very distinctive for being mostly composed of loose, adjustable, relocatable elements, including deck chairs, umbrellas, and sand. These beaches generally offer a large number and diversity of free seating options that visitors can rearrange to support a wide variety of uses (talking in groups, resting, sunbathing, dining) and users (children, families, the elderly). Sand is a loose and highly engaging landscape material that people can freely adjust to make themselves comfortable, and which invites playful transformation. Visitors to artificial beaches almost always modify the setting, dragging deck chairs, umbrellas, and side tables into clusters and turning them to optimize sun exposure, views, and privacy. Adults dig their feet into the sand and children build castles and draw patterns in it. At the end of each day, beach managers rake the sand smooth and remove all cigarette butts, resetting its pristine condition for inevitable further temporary modifications.

The Right to Disposition

This right enables someone to transfer the ownership of a space (and all the other rights that come with ownership) to others. Here it is important to distinguish between ownership of two kinds of property. The physical infrastructure of the artificial beach landscape is cheap and entirely mobile and thus set free from the politics of urban land. Some German temporary beaches have been relocated to serve new sites and new publics. The French public examples highlight bringing beaches to less affluent inland residents who don't otherwise have access to them. These transfers overcome the scarce and very unequal geography, accessibility, and property rights of real beaches. In this way, artificial beaches can enable a very radical spatial politics: they mass produce and redistribute exotic leisure spaces, "freeing" the beach for everyone everywhere.

In terms of the disposition of real property, the underlying land where city beaches sit is usually only rented cheaply and for the short term from a commercial landowner or local government. Most of these beaches are designed to rapidly vacate their site if there is a more pressing public purpose (e.g., managing river flooding) or private purpose (e.g., if the empty site finds a new buyer or commercial tenant). Alternative sites are usually available. When the site of Düsseldorf's first city beach was developed into an international hotel, its operator relocated onto a moored boat in nearby Cologne. When Munich's *Kulturstrand* had too many complaints from nearby residents, they agreed with the city to rotate between several locations. The loosened relationship between land and place that these artificial beaches establish can help overcome the contentious property politics of natural beaches, which ideally should be equally available to all.

The Dynamism of the Artificial Beach

The varying politics of these city beaches result from the diversity of new actors, locations, time frames, funding, management approaches, and programming that have produced them. New contexts and new ideas also keep transforming the relationships between these various actors.[30] These innovations can create new conflicts, but also new opportunities. If politics is the art of the possible, artificial beaches are creative places that expand what the politics of the beach can be and how they can be managed. These beaches are designed to resemble a natural space of escape and enjoyment, but their every element, every grain of sand, is shaped by human motivations and actions. There are no fixed lines in this sand.

Artificial beaches can significantly shape and reshape the wider politics of public space. This is especially the case for government- and community-run artificial beaches, which reallocate leisure spaces, rights, and pleasures to social groups that lack them. City beaches that are privately financed and operated, like other more durable forms of privately owned public spaces, also change the provision and scope of public open space amenity. They physically extend the public realm, transforming various types of privately owned and previously inaccessible sites into spaces that are available and enjoyable for broad publics. One particularly distinctive example is the numerous artificial beaches developed on the rooftops of downtown parking garages in Germany:[31] "City beaches on the roofs of parking lots and department stores are probably the most striking examples of this (gradual?) spatial and temporal unanchoring—the 'beach' experience is possible anytime and anywhere."[32]

City beach entrepreneurs take risks that expand ideas about where and what public space can be, and they provide the capital, labor, and know-how to deliver it. What is most important about these particular privately owned public spaces, in political terms, is their flexibility. One artificial city beach in the Netherlands shows that their politics can even remain dynamic in the context of large-scale, masterplanned, neoliberal property development and place marketing. The beach *Blijburg an See* was created on an artificial, dredged-sand island within the polderized IJmeer, with the support of the Amsterdam city government's public development corporation, to attract potential apartment purchasers with the temporary enjoyment of a day visit to the seaside.[33] As the phased housing development progressed, the beach, programmed with live music and events, grew in popularity, and its operator relocated it several times to yet undeveloped parcels further along the sandy island. Eventually, following petitions from a growing resident constituency, the masterplan was revised to include the beach as a permanent feature of the new neighborhood, and other infrastructure had to be replanned around it. This artificial city beach developed its own public and shifted the goals and rules of the urban planners.

It is generally difficult to adjust the politics of a housing development or shopping arcade once they have been built; their regimes of spatial rights are literally concrete. Artificial beaches demonstrate one mode of developing, managing, and using urban open space which exemplifies very small-scale creativity, flexibility, playfulness, and openness. Governments and communities can do political and manual work to adjust these places and make them as equitable, useful, beautiful, and profitable as they wish. As the community leaders of Vaihingen's *Strandleben* noted, "I think that this is the secret to success. No one simply set it up. One had to fight for it and one has to do something for this beach every day."

City beaches are complex, dynamic milieux where a broad range of actors constantly reframe the rights, benefits, and opportunities of urban spaces. Local governments use them to promote social inclusion as well as city marketing for economic and property development. City beach entrepreneurs are motivated not only by profits, but also by the desire to support neglected neighborhoods, support cultural activities, and change their own lifestyle. Users also exercise considerable freedom of action, appropriation, and modification within these landscapes.

These new beaches give people new places to relax, and they also relax the rules of urban spaces and social interaction, helping to playfully reimagine the right to the city.

ACKNOWLEDGMENTS

This research was supported by a grant from the Australian Research Council (Project DP180102964), funded by the Australian Government, and by a Senior Research Fellowship from the Alexander von Humboldt Foundation, Germany. Thanks to Patricia Aelbrecht, Birgit Billinger, Dagmar Meyer-Stevens, and Djamila Vilcsko for research support.

NOTES

1 McKenzie Wark, *The Beach Beneath the Street: The Everyday Life and Glorious Times of the Situationist International* (London: Verso, 2011).

2 Wark, *Beach Beneath the Street.*

3 Antoine Canivez, Deputy Director of Public Services, City of St. Quentin, personal interview, August 11, 2010.

4 Bertrand Delanoë, quoted in Michèle De La Pradelle and Emmanuelle Lallement, "Paris Plage: 'The City is Ours,'" *Annals of the American Academy of Political and Social Science* 595, no. 1 (2004): 135.

5 Mairie de Paris, *Paris Plages: Événement* (Paris: Mairie de Paris, 2007).

6 John Lichfield, "A Day at the Beach—without Leaving Paris's Violent Estates," *Independent*, July 11, 2005, www.independent.co.uk.

7 Christina Passariello, "French Beaches Reflect Divide Between Affluent, Immigrants," *Wall Street Journal*, August 7, 2007, www.wsj.com.

8 Sabine Gassner, Ulrike Schmidt-Hitschler and Thomas Hitschler, Organisers of *Strandleben*, Vaihingen an der Enz, personal interview, August 6, 2010.

9 Gassner et al., interview.

10 Marcus Benz, organiser of *Esslingen Stadtstrand*, personal interview, July 29, 2010.

11 Stadt Schwedt/Oder, "Klettergarten-Einweihung mit Strandbar," accessed April 16, 2012, www.schwedt.eu.

12 Benjamin David, manager of *Kulturstrand*, Munich, personal interview, July 22, 2010.

13 Quentin Stevens, "Throwntogether Spaces: Disassembling 'Urban Beaches,'" in *The Routledge Companion to Public Space*, ed. Vikas Mehta and Danilo Palazzo, (Abingdon, UK: Routledge, 2020); Klaus Sachs, "Stadtstrände—empirische Annäherungen an postmoderne Landschaften," 2010, accessed March 24, 2021, www. geog.uni-heidelberg.de.

14 Quentin Stevens and Mhairi Ambler, "Europe's City Beaches as Post-Fordist Placemaking," *Journal of Urban Design* 15, 4 (2010): 515–37.

15 Quentin Stevens, "City Beaches: Enlivening Marginal Spaces in Germany," in *Routledge Handbook of Urban Public Space*, ed. Karen A. Franck and Te-Sheng Huang (New York: Routledge, 2023).

16 Stevens and Ambler, "Europe's City Beaches."

17 Quentin Stevens, "Artificial Waterfronts," *Urban Design International*, 14 (2009): 3–21.

18 Quentin Stevens, "Sandpit Urbanism," in *Enterprising Initiatives in the Experience Economy*, ed. Britta Knudsen, Dorthe Christensen, and Per Blenker (New York: Routledge, 2015).

19 Gassner et al., interview.

20 Stevens, "Throwntogether Spaces."

21 Quentin Stevens, "Temporariness Takes Command: How Temporary Urbanism Re-Assembles the City," in *Transforming Cities Through Temporary Urbanism*, ed. Lauren Andres and Amy Zhang (Cham, Switzerland: Springer, 2020).

22 Philipp Oswalt, Klaus Overmeyer, and Philipp Misselwitz, *Urban Catalyst: The Power of Temporary Use* (Berlin: DOM publishers, 2013).

23 Senatsverwaltung für Stadtentwicklung Berlin, ed., *Urban Pioneers: Temporary Use and Urban Development in Berlin* (Berlin: Jovis, 2007).

24 Stevens, "Throwntogether Spaces."

25 Kevin Lynch, *Good City Form*, (Cambridge, MA: MIT Press, 1981), 230; Quentin Stevens, "Kevin Lynch," in *Key Thinkers on Cities*, ed. Alan Latham and Regan Koch (New York: Routledge, 2017).

26 Michael "Carlos" Carl, manager, *Strand 22*, Jena, personal interview, June 25, 2010.

27 Stefanie Kahls, *Stadtstrände: Südseefeeling in deutschen Großstädten* (Hamburg: Diplomica, 2009).

28 Ray Oldenburg, *The Great Good Place* (New York: Paragon House, 1989).

29 Sachs, "Stadtstrände."

30 Quentin Stevens and Kim Dovey, *Temporary and Tactical Urbanism: (Re)Assembling Urban Space* (New York: Routledge, 2023).

31 Stevens, "City Beaches."

32 Sachs, "Stadtstrände," 7.

33 Stevens and Ambler, "Europe's City Beaches."

ACKNOWLEDGMENTS

SETHA LOW

I want to begin by thanking Dr. Suzanne Scheld and Dr. Dana Taplin who share my passion for the study of beaches and the politics that animate them. Suzanne and Dana are my coauthors of *Rethinking Urban Parks: Public Space and Cultural Diversity*, our 2005 book based on park ethnographies that includes chapters on Orchard Beach, in the Bronx, and Jacob Riis Beach in Queens, New York City. We collaborated on fieldwork at Jones Beach and the shoreline of Lake Welch as part of a New York State Park study in 2012 reported in *Why Public Space Matters*. Recently we completed research in twenty-two New York State parks, some of which contained waterfront walkways and beaches. I am grateful for their insights and collegiality. I would also like to thank the Open Space Institute for their support and funding of the New York State parks projects.

In spring 2023, I taught a graduate seminar on Beach Politics and had the pleasure of getting to know a group of dedicated and talented graduate students from anthropology, geography, environmental psychology, art history, urban education, and liberal studies. I would like to thank them— Anuja Mukherjee, Mara Lasky, Jennifer Jones, Dinorah Hudson, Simone Cecilia Parker, Neyshka Diaz, Jesse McNeill, and Anna Schlenz—for their informative discussions of their own fieldwork and insights into the readings and theories of beach history and ethnography. Anna wrote about the beach restrictions and local practices in Narragansett, Rhode Island, and shared her research findings so that I could undertake my own site visit. These graduate students and the exciting seminar contributed to the success of this project.

Colleagues also offered their support by reading sections of the manuscript, providing technical support, and editing. Four of the contributors—Dr. Katherine McCaffrey, Dr. Paul Rouse, Dr. Matilde Córdoba Azcárate, and Dr. Dana Taplin—edited chapters, and contributor Dr. Bryce

DuBois created the author abstract and keyword list. An anthropology colleague and former graduate student, Dr. Charles Price, spent hours listening to my initial framing of the book and going over the chapters to identify the threads that would best organize it. It was his idea to focus on the rationales and justifications for beach exclusion. Dr. Stephanie Kane shared her expertise on beaches as geophysical formations and guided me in highlighting what this volume accomplishes. Dr. Melissa Checker saved me at the very end of the manuscript submission process by editing the introduction and giving it clarity and a bit of punch. I am grateful to these colleagues for their friendship and contributions to the completion of this book.

I want to particularly thank Jennifer Hammer at New York University Press for her support and encouragement. Jennifer is the rare combination of a wonderful person and a good critic. She knows how to help when help is needed, but she also knows how to hold the line and let you know when something is not working and needs to be attended to. In this editing project, Jennifer took on the role of a coeditor promising to offer an eye and an ear if I lost my way. She was the person I called when unsure of what direction to take and she guided the writing process from conceptualization through manuscript delivery.

Another key person who helped hold this project together was Melissa Matthews. Melissa is an editor usually involved in the book development and copyediting process, but when I realized that I would not be able to format all the chapters and organize the manuscript she expertly stepped in. Melissa is also an anthropologist who had just completed her master's thesis. Her background and accomplishments made her the perfect person to help me on my way. I could not have delivered this manuscript without her.

Publication management and copyediting conclude the bookmaking process and can be the hardest and most complicated part. Sometimes there is no time or space to thank the people involved. Therefore, I want to take a moment to highlight and thank Ainee Jeong, the production associate and manager of copyediting and proofing. She was a miracle worker who sent clear instructions for the authors and guided me through dealing with multiple authors, illustrations, and etchings. Ann Boisvert, the copyeditor, also contributed an incredible job of correcting the many styles and formats submitted. I am indebted to both for their attention to detail and support that brought this volume to completion.

I dedicate this book to my late mother and to my sister, Anna Harwin, a staunch cheerleader especially on days when I thought the volume would never come together. Anna's sense of the importance of the project made it possible to keep working even when I had my doubts.

My husband Joel Lefkowitz has been my constant companion through-out my career. I want to thank Joel for his many kinds of support: editing the introduction and preface, revisiting the table of contents repeatedly, shopping and cooking dinner, and, most importantly, talking endlessly about the thousands of questions that arise—emotional and intellectual—throughout the writing and publishing process. His care makes it possible to take on large project such as this one.

ABOUT THE CONTRIBUTORS

MATILDE CÓRDOBA AZCÁRATE is Associate Professor of Communication at the University of California, San Diego. She is the author of *Stuck with Tourism: Space, Power, and Labor in Contemporary Yucatán* (University of California Press, 2020) and coeditor of *Tourism Geopolitics: Assemblages of Infrastructure, Affect, and Imagination* (University of Alabama Press, 2020).

CHARIS CHRISTODOULOU is Associate Professor of urban planning and design and Coordinating Member of the Research Unit for South European Cities at Aristotle University of Thessaloniki, School of Architecture. Her current research is focused on community-centered planning of heritage landscapes and tourism. She is the author of *Landscapes of Sprawl: Urbanization and Urban Planning. The Periphery of Thessaloniki* (University Studio Press, 2015) and coauthor of *Sustainable Urban Design? Basica Issues and Processes. A Handbook* (Kallipos, 2024).

BRYCE DUBOIS is Associate Professor and the Marine Policy and Management Program Coordinator in the Biology and Environmental Science Department at the University of New Haven. His work is focused on racialized coastal landscapes and the politics of ecological knowledge and praxis in order to develop more just approaches to climate adaptation and environmental governance. He has published on these topics in peer-reviewed journals including *Ecopsychology*, *Urban Ecosystems*, *Sustainability Science*, and *Urban Forestry & Urban Greening*.

KEVIN DURRHEIM is Distinguished Professor in Psychology at the University of Johannesburg. His coauthored and coedited books include *The Routledge International Handbook of Discrimination, Prejudice and Stereotyping* (Routledge, 2021), *Qualitative Studies of Silence* (Cambridge

University Press, 2019), *Race Trouble* (Lexington Books, 2011), and *Racial Encounter* (Taylor & Francis, 2005).

LEIGH GRAHAM is Senior Advisor for Innovation Research at the Bloomberg Center for Public Innovation at Johns Hopkins University. She is an expert on racial and neighborhood equity in urban resilience politics and works to bridge research and practice to improve public sector innovation. She has published work across a wide range of policy domains in peer-reviewed journals such as the *Journal of the American Planning Association, Global Environmental Change,* and the *Journal of Public Health Management and Practice.*

NATASHA HOWARD is an Assistant Professor in Geography and Environmental Studies and Africana Studies at the University of New Mexico. Her work focuses on race and geography, and examines anti-Black spatial politics in the United States and Latin America. Her most recent work includes an article in the *Journal of Pan African Studies,* "Spatializing Blackness in New Mexico," and a chapter, "The Reproduction of the Anti-Black Misogynist Apparatus in U.S. and Latin American Pop Culture," in the anthology *Black Women and Social Justice Education* (SUNY, 2019).

KURT IVESON is Professor of Urban Geography at the University of Sydney. His current research is focused on the governance of the outdoor media landscape and the spatial politics of urban informatics systems. He is the coauthor of *Everyday Equalities: Making Multicultures in Settler Colonial Cities* (University of Minnesota Press, 2019).

ANDREW W. KAHRL is Professor of History and African American Studies at the University of Virginia. He specializes in the social, political, and environmental history of race and real estate in the twentieth-century United States. He is the author of *The Land Was Ours: How Black Beaches Became White Wealth* (University of North Carolina Press, 2016), *Free the Beaches: The Story of Ned Coll and the Battle for America's Most Exclusive Shoreline* (Yale University Press, 2018), and *The Black Tax: 150 Years of Theft, Exploitation, and Dispossession in America* (University of Chicago Press, 2024).

NADINE KHAYAT is a Senior Lecturer at the School of Architecture and Design at the American University of Beirut. Her work is focused on migration and displacement, multiculturalism, leisure practice, and physical and social integration across public space typologies in the Global South. Most recently she published *Exploring Urban Co-presence and Migrant Integration on Beirut's Seafront* with Clare Rishbeth (Migration Studies, 2023).

SABINE KNIERBEIN holds a PhD in European Urban Studies and an Internationale Urbanistik PD, and is Professor at the Interdisciplinary Centre for Urban Culture and Public Space at TU Wien, Austria. She explores everyday life and urbanization; the urban political, and democracy; disruptive precarity, and crises; intersectional methodology. She is coeditor of *Public Space Unbound: Urban Emancipation and the Post-Political Condition* (2018), *Care and the City: Encounters with Urban Studies* (2022), and *Unsettled Urban Space: Routines, Temporalities, and Contestations* (2023), all with Routledge.

SETHA LOW is Distinguished Professor of Environmental Psychology, Geography, Anthropology, and Women's Studies, and Director of the Public Space Research Group at the Graduate Center, City University of New York. Her current research is on the difficulties of creating collectivity and collective governance in the United States. She is author of *Spatializing Culture: The Ethnography of Space and Place* (Routledge, 2017), *Why Public Space Matters* (Oxford University Press, 2023) and, with Mark Maguire, *Trapped: Life Under Security Capitalism and How to Escape It* (Stanford University Press, 2024).

KATHERINE T. MCCAFFREY is Professor of Anthropology at Montclair State University. Her research focuses on militarism, migration, and displacement. She is author of *Military Power and Popular Protest: The US Navy in Vieques, Puerto Rico* (Rutgers 2002) and is finalizing a book about refugee resettlement in New Jersey.

MARIANO D. PERELMAN holds a PhD in Anthropology (Universidad de Buenos Aires), is a researcher of CONICET, and is Professor of the Department of Anthropology at the Universidad de Buenos Aires. His work is focused on how people deal with uncertainties in context of crisis. Most recently, he published *Coping with Inflation, Living in Crisis, Thinking about the Dollar: Ways of Life in Buenos Aires* (Current History, 2024), *Informal Collection in Buenos Aires. Behind and Beyond the Crisis* (Vibrant, 2023), *La venta ambulante en Buenos Aires: Economía(s) (i)legales, estética y circulación de objetos* (Etnográfica, 2022), and, with M. Di Virgilio, *Desigualdades urbanas en tiempos de crisis* (University of Nebraska Press, 2021).

KEISHA-KHAN Y. PERRY is Presidential Penn Compact Associate Professor of Africana Studies at the University of Pennsylvania. Her research is focused on race, gender, and politics in the Americas; urban geography; questions of citizenship; intellectual history and disciplinary formation; and the interrelationship between scholarship, pedagogy, and political engagement. She is the author of *Black Women against the Land Grab: The Fight for Racial Justice in Brazil* (University of Minnesota Press, 2013).

CLARE RISHBETH is Professor of Inclusive Landscapes in the Department of Landscape Architecture, University of Sheffield, UK. Her research focuses on intercultural urban spaces, nature connections, and migration, often using participatory methodologies. In 2023–24 she was appointed as an Arts and Humanities Research Council Innovation Scholar.

PAUL ROUSE is Professor of History at University College Dublin. He has written extensively on the history of sport and is author of *Sport and Ireland: A History* (Oxford University Press, 2015). He also teaches and writes on popular culture, public spaces, the history of agriculture and Irish history in general.

BENJAMIN HEIM SHEPARD is Professor of Human Service at New York School of Technology, City University of New York. His published works include *Sustainable Urbanism and Direct Action: Case Studies in Dialectical Activism* (Roman and Littlefield, 2021), *Illuminations on Market Street: A Story about Sex and Estrangement, AIDS and Loss, and Other Preoccupations in San Francisco* (Columbia University Press,

2019), and *White Nights and Ascending Shadows: An Oral History of the San Francisco AIDS Epidemic* (Cassell, 1997).

QUENTIN STEVENS is Professor of Urban Design at RMIT University in Melbourne. His research examines people's perceptions and uses of public spaces and public artworks and how they are designed, planned, procured and managed. His recent books include *Temporary and Tactical Urbanism* with Kim Dovey (Routledge, 2022) and *Activating Urban Waterfronts* (Routledge, 2020).

DANA TAPLIN manages land use and environmental review procedures for urban stormwater management projects at the New York City Department of Environmental Protection. Dr. Taplin is a coauthor of *Rethinking Urban Parks: Public Space and Cultural Diversity* (University of Texas Press, 2005) and coauthor of a number of articles on rapid ethnographic procedures in scholarly journals and books, including "Conservation and the People's Views," in the anthology *Human Centered Built Heritage Conservation* (Taylor & Francis, 2019), and "Rapid Ethnographic Assessment in Urban Parks, a Case Study of Independence National Historical Park," in *Human Organization* 61, no. 1, (2002).

ANA VILA-CONCEJO is Professor of Coastal Geomorphology at the University of Sydney and deputy director of One Tree Island Research Station. Her research is focused on the contemporary processes and morpho-dynamics of coastal systems. Her work has been published in *Estuaries and Coasts, Geomorphology*, and the *Journal of Coastal Research*.

INDEX

Page numbers in *italics* indicate figures.

Aboriginal peoples, 125

activism, 13–14, 47; academia and, 76; of Black women, 236–38; in Brazil, 236; community, 163; Dollymount Strand and, 274; ethnography and, 161; privatization and, 111; public space, 74–75; in Puerto Rico, 251; urban, 161. *See also* protests

Adams, Douglas, 172

adaptation, private property and, 134

African Americans: displacement of, 181–83; land ownership in Brazil of, 239; land ownership of, 7; public housing and, 190

African National Congress, 217

Afro-Paradise (Smith, C.), 232

Alan (hurricane), 229

Albuquerque, New Mexico, 12; beaches of, 196–97, *198–99*, 202, 204–5; Black residents of, 197, 200, 204–7, 210n12; Conservancy Beach, 196; discrimination in, 204–5; Hispanic population of, 201; home ownership in, 205; housing in, 204, 206–7; institutionalizing collective memory of, 205–9; Open Space System, 200, 202, *203*; politics of memory in, 201–3; privatization in, 207; racism in, 207; recreational space of, 206–7, 209; segregation in, 204, 207; sites of memory in, 200–201; South Broadway, 204, 206; sunbathing in, 209; tourism and, 209; wildlife and

environment of, 202. *See also* Tingley Beach

Albuquerque Biological Park (ABQ BioPark), 196

Albuquerque Museum, 197

aliveness, 237

American University (Beirut), 70

Anti-Defamation League, 26

"El Apagón-Aquí Vive Gente," 243, 244

apartheid, 212, 215, 217, 219–20

Arab culture, 67

Arab Spring, 67

Ardilaun, Lord, 268, 269

Argentina: beach time in, 82–83, 86–88, 92–94; football and violence in, 95n5; labor in, 82, 93; laws in, 82; space governance in, 84; *tetazo* protests of, 83, 88–94; vending in, 82–88; work and commerce in, 83–88, 94n2

artificial beaches. *See* city beaches

Arverne, 182, 183, 190, 192

at-risk properties, 128

Attersee (Austria), 104–7, *107*

Australia. *See* Sydney, Australia

Austria, 97; Attersee, 104–7, *107*; carbon footprint of, 108; constitution of, 109; depoliticization of public space in, 104–7; forests of, 105; lakes of, 105–6, 109; neoliberalism in, 110; OBf, 105–6; privatization in, 99, 104, 106, 108–9; property rights in, 108–9; public access in, 98, 101, 111; regulations in, 108–9;